T0339597

HANDBOOK FOR TRANSVERSELY FINNED TUBE HEAT EXCHANGER DESIGN

HANDBOOK FOR TRANSVERSELY FINNED TUBE HEAT EXCHANGER DESIGN

EUGENE PIS'MENNYI

GEORGIY POLUPAN

IGNACIO CARVAJAL-MARISCAL

FLORENCIO SANCHEZ-SILVA

IGOR PIORO

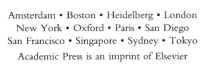

Amsterdam • Boston • Heidelberg • London
New York • Oxford • Paris • San Diego
San Francisco • Singapore • Sydney • Tokyo
Academic Press is an imprint of Elsevier

Notices
Knowledge and best practice in this field are constantly changing. As new research and
experience broaden our understanding, changes in research methods, professional practices,
or medical treatment may become necessary.

Practitioners and researchers must always rely on their own experience and knowledge in
evaluating and using any information, methods, compounds, or experiments described herein.
In using such information or methods they should be mindful of their own safety and the
safety of others, including parties for whom they have a professional responsibility.

To the fullest extent of the law, neither the Publisher nor the authors, contributors, or editors,
assume any liability for any injury and/or damage to persons or property as a matter of
products liability, negligence or otherwise, or from any use or operation of any methods,
products, instructions, or ideas contained in the material herein.

British Library Cataloguing-in-Publication Data
A catalogue record for this book is available from the British Library

Library of Congress Cataloging-in-Publication Data
A catalog record for this book is available from the Library of Congress

ISBN: 978-0-12-804397-4

For information on all Academic Press publications
visit our website at https://www.elsevier.com/

Working together
to grow libraries in
developing countries
www.elsevier.com • www.bookaid.org

Publisher: Joe Hayton
Senior Acquisition Editor: Lisa Reading
Editorial Project Manager: Peter Jardim
Production Project Manager: Kiruthika Govindaraju
Designer: Victoria Pearson Esser

Typeset by TNQ Books and Journals

CONTENTS

ABOUT THE AUTHORS

Eugene Pis'mennyi graduated as a mechanical engineer within the area of steam-
generator design from the National Technical
University of Ukraine, Kyiv Polytechnic
Institute. He obtained his Ph.D. within the
area of thermal physics at the same institution.
Dr. Pis'mennyi earned a degree of Doctor of
Technical Sciences from the Institute of Engineering Thermophysics, National Academy
of Sciences of Ukraine. Dr. Pis'mennyi is an
internationally recognized scientist within the areas of nuclear engineering
and heat transfer. He is an author of more than 400 publications, including
eight technical books, 16 patents, and more than 50 technical reports. He has
supervised 15 Ph.D. and more than 100 master in science and bachelor's
theses. Currently, Dr. Pis'mennyi is a dean of the Heat Power Engineering
Faculty and the Chief of Nuclear Power Plants and Engineering Thermo-
physics Department at the National Technical University of Ukraine, Kyiv
Polytechnic Institute.

His research interest is convective heat transfer and methods of its
enhancement.

His contact e-mail: evgnik@i.com.ua.

Georgiy Polupan graduated as a mechanical
engineer within the area of steam generator
design from the National Technical University of Ukraine, Kyiv Polytechnic Institute.
He obtained his Ph.D. within the area of
thermal physics at the same institution.

His major area of study is thermal processes in steam generators, thermo-physics,
and efficient use of energy. He is an author
of more than 300 publications, including
three technical books, two patents, and 67 papers in refereed journals.

G. Polupan was a researcher at the Thermal Physics Department from
1972 to 1985 and a professor of the Thermal Engineering Faculty of the

National Technical University of Ukraine, Kyiv Polytechnic Institute from 1985 to 1999. Since 1999, he has been a full professor at the National Polytechnic Institute of Mexico, Superior School of Mechanical and Electrical Engineering, Thermal Engineering and Applied Hydraulic Laboratory.

His contact e-mail: gpolupan@ipn.mx.

Ignacio Carvajal-Mariscal graduated as a mechanical engineer from the Moscow Power Engineering Institute (Russia) in 1994. He obtained an M.Sc. and Ph.D. in thermal engineering from the same institution in 1996 and 1999, respectively. He was a visiting professor at the Mechanical and Mechatronics Engineering Department, University of Waterloo, Waterloo, Canada, from 2008 to 2009. His major area of study is convective heat transfer. He is an author of more than 200 publications, including one technical book and 23 papers in refereed journals.

Currently, Dr. Carvajal-Mariscal is a full professor at the National Polytechnic Institute of Mexico, Superior School of Mechanical and Electrical Engineering, Thermal Engineering and Applied Hydraulic Laboratory. His research interests are heat-transfer enhancement, experimental fluid dynamics, and heat-exchanger design.

Dr. Carvajal-Mariscal has been an ASME member since 2008.

His contact e-mail: icarvajal@ipn.mx.

Florencio Sanchez-Silva graduated as a Mechanical Engineer from the National Polytechnic Institute of Mexico, Mexico City, in 1974. He obtained an M.Sc. in thermodynamics and Ph.D. in thermal engineering from the National Superior School of Mechanics and Aerothermics of Poitiers, France, in 1980. He was a post doctorate in two-phase flow dynamics at the University of Pisa, Italy, in 1988−1989 and spent his sabbatical leave at Texas A&M University from 2000 to 2001. His major area of study is two-phase flow dynamics and phase change heat transfer.

Currently, he is a full professor at the National Polytechnic Institute of Mexico, Superior School of Mechanical and Electrical Engineering, Thermal Engineering and Applied Hydraulic Laboratory. His research interests are two-phase flow dynamics, heat pipes, energy saving, and development and analysis of thermal energy systems.

To date, Dr. Sánchez-Silva has 38 refereed journal publications, 250 refereed papers in conference proceedings, 15 technical reports, two technical books, and three chapters in technical books. He has supervised three Ph.D., 44 master in science, and 31 bachelor's theses on two-phase flow, heat transfer, fluid mechanics, metrology, and thermal systems topics.

Dr. Sanchez-Silva is a Fellow of the Societé des Ingenieurs et Scientifiques de France, Mexican Section (since 1987) and has been a member of the Mexican Academy of Sciences since 2004.

His contact e-mail: fsanchezs@ipn.mx.

Professor Igor Pioro, Ph.D. (1983), Doctor of Technical Sciences (1992), professional engineer (2008), Fellow of the American Society of Mechanical Engineers (2012), Fellow of the Engineering Institute of Canada (EIC) (2013), and Fellow of the Canadian Society of Mechanical Engineers (2015), is an internationally recognized scientist within the areas of nuclear engineering and heat transfer. He has authored/coauthored more than 425 publications, including nine technical books, 15 chapters in encyclopedias,  handbooks, and technical books, 88 papers in refereed journals, 241 papers in refereed proceedings of international and national conferences and symposiums, 26 patents and inventions, and 46 major technical reports.

Dr. Pioro graduated from the National Technical University of Ukraine, Kiev Polytechnic Institute with an M.A.Sc. in thermal physics in 1979. After that, he worked in various positions, including engineer, senior scientist, deputy director, professor, director of the graduate program in nuclear engineering, and associate dean. Currently, he is associated with the Faculty of Energy Systems and Nuclear Science University of Ontario Institute of Technology (Oshawa, Ontario, Canada).

Dr. Pioro is a Founding Editor of the ASME Journal of Nuclear Engineering and Radiation Science (from 2014). He was a Chair of the Executive Committee of the Nuclear Engineering Division of the ASME (2011–2012) and a Chair of the International Conference On Nuclear Engineering (ICONE20-POWER2012).

Professor Pioro has received many international and national awards and certificates of appreciation including Service Recognition Award from the ASME (2014), Honorary Doctor's degree from the National Technical University of Ukraine, Kiev Polytechnic Institute (2013), The Canadian Nuclear Society (CNS) Education and Communication Award (2011), and the ICONE Award from the ASME (2009), among others.

FOREWORD

It is a pleasure for me to respond to the invitation of writing a foreword to the English translation of the original book *Manual Para el Cálculo de Intercambiadores de Calor y Bancos de Tubos Aleteados*. In my opinion, this is a very complete and practical handbook on the topic, both for advanced students, researchers, and especially for professional practitioners in mechanical, energy, chemical, and thermal engineering. It covers exhaustively the dimensioning and thermo-fluid performance calculation of banks of tubes as a part of the layout of industrial heat exchangers, incorporating transversal fins as external area extenders to compensate the low heat carrying capacity of flowing gases and among them, air and combustion fumes.

It represents an excellent example of international collaboration on technological progress between the Faculty of Thermal Engineering of the National Technical University of Ukraine, Kiev Polytechnic Institute and the Superior School of Mechanical and Electrical Engineering, Thermal Engineering and Applied Hydraulic Laboratory of the National Polytechnic Institute in México. It also represents an example of the high-quality work of the laboratories where the tests were performed. The modern methodology chosen has been extensively communicated to the scientific and technology communities and includes empirical information that gathers practical experience.

The handbook deals with the internal-flow convection coefficient for smooth round tubes or tubes that could be treated as equivalently round, with the conduction of the wall and with the external convection coefficient for circular, square, and helical fins, considered as components of the overall heat transfer coefficient, including the effect of particle-containing gases. Algorithms for both the pressure loss of the internal and external flows are offered in the book, and it even provides a mechanical-strength calculation procedure for the wall thickness under thermal stresses and internal pressure. The book introduces correlations valid for a wide range of the influencing parameters under steady-state operational conditions without forgetting the influence of boiling and radiation. Care has been taken to include real-world effects such as flow maldistribution and flow nonuniformities. The methodology relies on the use of nondimensional groups, which guarantees consistency and wide applicability.

A fully solved exercise is offered at the end of the book, illustrating the top-down approach that is specially oriented toward power plant heat exchangers.

The Spanish edition of the book has obtained an excellent acceptance, especially in Latin America and Spain. Its translation to the English will broaden the community of professionals that can use it.

Antonio Lecuona Neumann
Professor of the Department of Thermal and Fluids Engineering,
ITEA group
Universidad Carlos III de Madrid, Spain
January 2012

PREFACE

The present handbook contains a single procedure of calculating heat transfer and aerodynamic resistance of convective heating surfaces fabricated in the form of tube bundles with transverse circular, square, and helical fins. The procedure is based on results of extensive experimental studies conducted at the Department of Nuclear Power Plants and Engineering Thermal Physics of the National Technical University of Ukraine, Kiev Polytechnic Institute [1−3] in collaboration with researchers of Superior School of Mechanical and Electrical Engineering, Thermal Engineering and Applied Hydraulic Laboratory of the National Polytechnic Institute in México. This procedure has been put through a thorough evaluation at leading scientific research and design organizations in Ukraine and Russia, and it has been recognized as the most accurate and universal in comparison with other methods used for thermal and aerodynamic calculations [4−6]. Based on that, it has been included in the currently operative normative materials of Ukraine and Russia.

A system of correlations for heat transfer and aerodynamic resistance, used in the present handbook, makes it possible to calculate, within the range of Reynolds numbers $\mathbf{Re} = 3 \times 10^3 - 2 \times 10^5$, heat transfer and resistance coefficients for staggered and in-line tube banks (bundles) alike with the fin coefficients of $\Psi_r = 1.2 - 39.0$ and the relative transverse σ_1 and longitudinal σ_2 spacings: $\sigma_1 = 1.7 - 6.5$ and $\sigma_2 = 1.3 - 9.5$ ($\sigma_1 / \sigma_2 = 0.3 - 5.2$). That is, it embraces all practical needs.

The procedure provides calculations for finned surfaces operating within conditions of clean and dust-laden flows alike, including finned convective heating surfaces of boilers [7].

The handbook also includes the procedure for calculating hydraulic resistance of tube bundles with respect to the internal heat-transfer medium [8,9], and the procedures for calculating the temperature mode of finned tubes [10] and for the strength design based on recommendations of the standard methods of hydraulic and thermal calculations of boilers and the standards of the strength design [11−13]. Examples of the calculations are also provided.

Thus, the offered publication represents a completed complex of methodical materials for calculating and designing convective surfaces, constructed from transversely finned tubes, and it is intended to be used by design engineers, scientists, and students of power engineering and power machine design specialties.

NOMENCLATURE

A	outside heat-transfer surface area, m^2
A_{in}	inside heat-transfer surface area, m^2
A_r	heat-transfer surface area of fins, m^2
A_{rct}	surface of finning–carrying tubes, m^2
A_{rs}	heat-transfer surface area of finned tube segments on outside, which is defined as difference between calculated heating surface A and surface of unfinned segments (bends and junction regions), m^2
A_t	tube part of heating surface not occupied by fins, m^2
C_r	coefficient in similarity equation for aerodynamic resistance
C_q	coefficient in similarity equation for heat transfer
C_z	correction for number of tube rows in bundle in direction of gas flow
c_p	specific heat at constant pressure, J/kg K
c_v	specific heat at constant volume, J/kg K
c_{sq}	side of square fin, m
$D = d + 2 \cdot l_r$	finning diameter, m
d	outside diameter of fin-carrying tube, m
d_{in}	inside diameter of tube, m
E	theoretical efficiency of fin
G_f	mass flow rate of internal heat-transfer medium, kg/s
G_g	mass flow rate of external heat-transfer medium, kg/s
h	heat transfer coefficient, W/m^2 K; specific enthalpy, J/kg
h_c	convective heat transfer coefficient, W/m^2 K
h_{1rdc}	reduced coefficient of heat transfer from outside (from gas), W/m^2 K
h_2	heat transfer coefficient from wall to internal medium, W/m^2 K
h_{fg}	latent heat of evaporation, J/kg
k	thermal conductivity of heat transfer medium, W/m K
k_t	thermal conductivity of tube metal, W/m K
k_r	thermal conductivity of fin metal, W/m K
$L_{c \cdot cr \cdot s}$	length of tube projection on calculated cross-section of gas conduit, m
L_{rs}	length of finned segment of tube, m
L_t	total length of heated unfinned tube segments, m

l_r	fin height, m
$l_{re} = l_r + \delta_2/2$	conventional fin height, m
P, p	pressure of heat-transfer medium, Pa
Q	heat transfer rate, W
R	thermal resistance, $m^2\,K/W$
$r_e = (D + \delta_2)/2$	conventional finning radius, m
$r_1 = d/2$	outside radius of tube, m
S_1	transverse spacing of tubes, m
S_2	longitudinal spacing of tubes, m
$S_2' = \sqrt{\frac{1}{4}S_1^2 + S_2^2}$	diagonal spacing of tubes, m
s	specific entropy, $J/kg\,K$
s_r	fin spacing, m
T, t	average temperature of internal medium in bundle, $^\circ C$
U	overall heat transfer coefficient, $W/m^2\,K$
u	velocity, m/s
v	specific volume, m^3/kg
z	number of tubes in bundle
z_{tcp}	number of tubes connected in parallel
z_1	number of tubes in transverse row of bundle
z_2	number of the tube rows in bundle in direction of gas flow

Greek Symbols

α	thermal diffusivity, m^2/s; $\left(\frac{k}{c_p \rho}\right)$
β	fin parameter, 1/m
β_{th}	volumetric thermal expansion coefficient, 1/K
Δ	difference
ΔP	aerodynamic resistance of tube bundles in crossflow, Pa
δ_r	average fin thickness, m
δ_t	nominal wall thickness of tube, m
δ_1	fin thickness at base, m
δ_2	fin thickness at tip, m
ε	contamination factor, $m^2\,K/W$
ζ_0	resistance coefficient of single tube row of bundle
μ	dynamic viscosity, Pa s
μ_r	factor taking account of fin widening toward base
Ψ	thermal efficiency factor
ρ	density, kg/m^3

$\sigma_1 = S_1/d$ relative transverse spacing of tubes
$\sigma_2 = S_2/d$ relative longitudinal spacing of tubes
ν kinematic viscosity, m^2/s
$\Psi_r = A_{rs}/A_t$ fin coefficient (the ratio of total outside surface of finning tubes to external surface of tube not occupied by fins)
ψ_E correction factor to theoretical efficiency of fin
ϑ average temperature of gas in bundle, °C
$\vartheta_{c\cdot r}$ gas temperature at entrance to calculated row, °C

Nondimensional Numbers

Nu   Nusselt number; $\left(\frac{hD}{k}\right)$
Pr   Prandtl number; $\left(\frac{\mu c_p}{k}\right) = \left(\frac{\nu}{\alpha}\right)$
Re   Reynolds number; $\left(\frac{uD}{\nu}\right) = \left(\frac{\rho u D}{\mu}\right)$

Subscripts

av   average
bnd   bend
col   collector
d   based on diameter
ent   entrance
f   fluid (liquid)
evap   evaporation
g   gas
h   heated
in   inside or inlet
int   intake
max   maximum
rad   radiation
t   tube
s   saturation
sup   supply
un   unheated
v   vapor
w   wall (at wall temperature)
$'$ at inlet
$''$ at outlet
$\overline{\square}$ parameter with overbar means average value

Acronyms

DC	Distributing collector
IC	Intake collector
SC	Supply collector
TRU	Turbo-refrigerating unit
WAHE	Water-to-air heat exchanger

CHAPTER 1

General Statements

To remain competitive, companies have to consider thermal energy savings and the reduction of cooling water consumption in their production processes. To achieve these goals, the implementation of heat recovery equipment and air-cooled heat exchangers is required.

It is well known that when the external fluid is a gas (air, natural gas, exhaust gases, etc.) in heat exchangers, low convection coefficients are obtained; this means that to increase heat exchange, extended surfaces should be used. These surfaces are fabricated as finned tube banks and become the core of gas—gas and gas—liquid type heat exchangers, such as heat recovery and air-cooled equipment.

The heat recovery equipment takes advantage of the thermal energy contained in the exhaust gases from combustion processes. In a combined cycle power plant, exhaust gas from gas turbines is used to generate steam in a heat recovery boiler. Other examples of this kind of equipment are the steam generator economizers and air preheaters that are widely used in industry. Thus, heat recovery increases the energy efficiency of thermal systems and reduces the emissions of greenhouse gases.

In recent decades, due to the increasing shortage of water for both human consumption and industrial uses, air-cooled heat exchange equipment is having increasing application. The aircooled heat exchangers are used in refineries, power plants, and for cooling products in chemical, food, and paper industries. The air-cooled condenser is one of the most used; they are installed in combined cycle power plants to condense the steam from one or more steam turbines.

Finned tubes are the heart of any gas—gas or gas—liquid type of heat exchanger. Finned tube banks are compact units of robust and corrosion-resistant construction. The type of finned tube is chosen (ie, the fin type and combination of materials) depending on the specific requirements of each process equipment unit. Commonly, tubes currently have circular or helical fins.

1.1 GEOMETRIC CHARACTERISTICS

Convective heating surfaces with circular, square, or helical fins represent tube bundles or tube banks with staggered or in-line arrangement of tubes in cross-flow. Bundles are fabricated from straight tubes, finned beforehand

Handbook for Transversely Finned Tube Heat Exchanger Design
ISBN 978-0-12-804397-4
http://dx.doi.org/10.1016/B978-0-12-804397-4.00001-X

through welding by high-frequency currents or in any other manufacturing process. Geometric characteristics of finned tubes are presented in Fig. 1.1.

Steps between tubes in in–line and staggered arrangements are shown in Fig. 1.2.

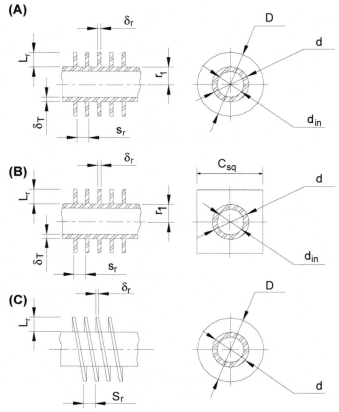

Figure 1.1 Geometric characteristics of finned tubes: (A) tube with circular fins; (B) tube with square fins; and (C) tube with helical fins.

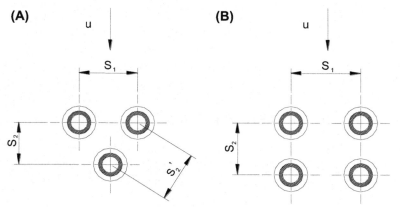

Figure 1.2 Spacing characteristics of bundles of finned tubes: (A) staggered arrangement; and (B) in-line arrangement.

CHAPTER 2

Heat-Transfer Calculations

2.1 BASIC EQUATIONS

2.1.1 Calculation of Quantity of Heat

The quantity of heat, taken up by the heating surface in unit time, is found from the equation:

$$Q = U \cdot A \cdot \Delta t \times 10^{-3}, \ \text{kW} \qquad [2.1]$$

where the overall heat-transfer coefficient U is determined according to Section 2.2 and the average temperature difference Δt according to Section 2.7.

2.1.2 Calculation of Heating Surface

The calculated heating surface A is taken to be equal to the total heating surface on the outside (on the side of gas) that comprises the heating surface of fins A_r and the heating surface of tubes A_t, excluding the tube surface occupied by fins,

$$A = A_r + A_t \qquad [2.2]$$

A_r for helical finning and circular fins is determined according to the expression:

$$A_r = \frac{\pi}{2}(D^2 - d^2 + 2 \cdot D \cdot \delta_r)\frac{L_{rs}}{S_r}z \qquad [2.3]$$

and for square fins,

$$A_r = 2\left(c_{sq}^2 - 0.785d^2 + 2 \cdot c_{sq} \cdot \delta_r\right)\frac{L_{rs}}{S_r}z \qquad [2.4]$$

where L_{rs} is the length of the finned segments of the tubes.

A_r is calculated from the equation:

$$A_r = \pi \cdot d\left[L_{rs}\left(1 - \frac{\delta_r}{S_r}\right)z + L_t\right] \qquad [2.5]$$

Here, L_t is the total length of the heated unfinned tube segments (bends and junction regions), which is determined from manufacturer's drawings.

Handbook for Transversely Finned Tube Heat Exchanger Design
ISBN 978-0-12-804397-4
http://dx.doi.org/10.1016/B978-0-12-804397-4.00002-1

2.2 OVERALL HEAT-TRANSFER COEFFICIENT

2.2.1 Calculation of Overall Heat-Transfer Coefficient

The overall heat-transfer coefficient of finned tubes can be calculated with sufficient accuracy as for a plane finned wall. Here, in the case of the tube bundle washed by the flow of a pure heat-transfer medium, as well as by the combustion products of gas or black oil, use can be made of the equation:

$$U = \frac{\Psi}{\frac{A}{A_{in}}\cdot\frac{1}{h_2} + \frac{A}{A_{in}}\cdot R_t + \frac{1}{h_{1rdc}}} \qquad [2.6]$$

where h_{1rdc} is the reduced heat-transfer coefficient, which is determined according to Section 2.3.1; h_2 is the coefficient of heat-transfer from the tube wall to the internal medium, which is determined according to Section 2.5. In calculations of the heat-transfer coefficients of steaming economizers and evaporation surfaces, the thermal resistance of heat transfer to the internal medium is generally disregarded, since in these cases $\frac{1}{h_{1rdc}} \ll \frac{1}{h_2}$; A is the total heating surface on the outside, which is determined from Eqs. [2.2] ÷ [2.5]; A_{in} is the inside heat-transfer surface, calculated from the equation:

$$A_{in} = \pi \cdot d_{int}(L_{rs}\cdot z + L_t) \qquad [2.6a]$$

Ψ is the thermal efficiency factor that accounts for a decrease in the heat absorption by the heating surface, resulting from its contamination and also from its nonuniform washing by gases, a partial overflowing of gases past it, and the formation of stagnation regions, and it is determined according to Section 2.6; R_t is the thermal resistance of the tube wall: for ordinary uniform tubes with welded or rolled-on finning and for cast finned tubes:

$$R_t = \frac{\delta_t}{k_t} \qquad [2.7]$$

The calculation of R_t for bimetallic finned tubes is shown:

$$R_t = \frac{\delta_t'}{k_t'} + R_{cont} + \frac{\delta_t''}{k_t''} \qquad [2.8]$$

Here; R_{cont} is the thermal contact resistance between the outside finned shell with the thickness δ_t' and the thermal conductivity k_t' and the inside smooth shell with the thickness δ_t'' and the thermal conductivity k_t''. R_{cont} is determined from the reference literature or experimentally. The thermal resistance of heat transfer from the outside $1/h_{1rdc}$ is generally much higher than the thermal resistance of the metal of the tube wall δ_t/k_t, and therefore the latter can be ignored ($\delta_t/k_t \approx 0$).

2.2.2 Calculation for Finned Tubes Exposed to Flow of Combustion Products of Solid Fuels

If the bundle of finned tubes is exposed to the flow of the combustion products of solid fuels, the heat-transfer coefficient is calculated from the equation:

$$U = \frac{1}{\frac{A}{A_{in}}\cdot\frac{1}{h_2} + \frac{A}{A_{in}}\cdot R_t + \frac{1}{h'_{1rdc}}} \qquad [2.9]$$

where h'_{1rdc} is determined according to Section 2.3.2 and the other quantities by Eq. [2.6].

2.3 REDUCED CONVECTIVE HEAT-TRANSFER COEFFICIENT h_{1rdc}

2.3.1 Calculation When Exposed to Flow of Clean Heat-Transfer Medium

Since the thickness of the radiative layer in the bundles of transversely finned tubes is small, the reduced coefficient of heat transfer from the outside is determined disregarding the inter-tube radiation.

In the case of the tube bundle exposed to a flow of a clean heat-transfer medium,

$$h_{1rdc} = \left(\frac{A_r}{A}E\cdot\mu_r\cdot\psi_E + \frac{A_t}{A}\right)h_c \qquad [2.10]$$

where h_c is determined according to Section 2.4, and A, A_r, and A_t according to Section 2.1.1.

The theoretical efficiency of the fin E for helical finning and circular fins is found as a function of the parameters $\beta\cdot l_r$ and D/d from the nomogram in Fig. 2.1, or for $\beta\cdot l_r \leq 2$ and $D/d \leq 3$, it is calculated from the equation:

$$E = \frac{\tanh\left(\beta l_r'\right)}{ml_r'} \qquad [2.11]$$

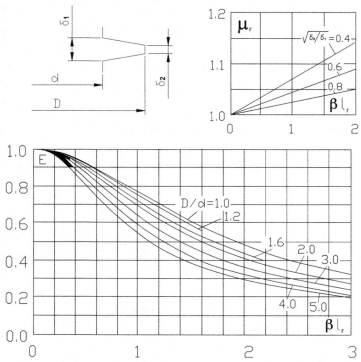

Figure 2.1 Fin correction factor μ_r, which takes into account fin widening toward base, and relations for theoretical efficiency of fin E.

where the following:

$$\tanh\left(ml'_r\right) = \frac{e^{2\beta l'_r} - 1}{e^{2\beta l'_r} + 1} \qquad [2.12]$$

$$\beta = \sqrt{\frac{2h_c}{\delta_r \cdot k_r}} \qquad [2.13]$$

The conventional fin height l'_r is determined using the relation:

$$l'_r = l_r\left[1 + \left(0.191 + 0.054 \times \frac{D}{d}\right) \times \ln\left(\frac{D}{d}\right)\right] \qquad [2.14]$$

The correction factor ψ_E to the theoretical efficiency of the fin E, which takes account of the nonuniformity of heat transfer over the fin surface, is calculated from the equation:

$$\psi_E = 1 - 0.016 \left(\frac{D}{d} - 1 \right) [1 + \tanh(2 \cdot \beta l_r - 1)] \qquad [2.15]$$

For fins with a trapezoidal cross section, the coefficient μ_r, which takes account the effect of the fin widening toward the root, is determined from the nomogram in Fig. 2.1 as a function of the parameters βl_r and $\sqrt{\delta_2/\delta_1}$ (δ_2 and δ_1 are the fin thicknesses at the tip and root).

In determining k_r, the average temperature of the fin metal is evaluated approximately using the equation:

$$T_r = \vartheta - (\vartheta - T) \cdot E' \qquad [2.16]$$

where ϑ and T are the average temperatures of the external and internal heat-transfer media. E' is the preliminary value given to the theoretical efficiency of the fin E considering its subsequent refinement.

For square fins, the theoretical efficiency of the fin E and the correction factor ψ_E are determined in exactly the same manner as for circular fins. Here, in the calculation relations, the following should be assumed:

$$D = 1.13 \cdot c_{sq} \qquad [2.17]$$

$$l_r = \frac{1.13 \cdot c_c - d}{2} \qquad [2.18]$$

2.3.2 Calculation When Heat-Transfer Medium Is From Combustion Products of Solid Fuels

When combustion products of solid fuels are used as a heat-transfer medium, the reduced coefficient of heat transfer from the outside of the heating surface is calculated with account for the contamination factor of the surface,

$$h'_{1rdc} = \left(\frac{A_r}{A} E \cdot \mu_r \cdot \psi_E + \frac{A_t}{A} \right) \cdot \frac{h_c}{1 + \varepsilon \cdot h_c} \qquad [2.19]$$

where the contamination factor ε, representing the thermal resistance of the layer of external deposits, is determined according to Section 2.6, and the other quantities are obtained from Eq. [2.10]. As with a clean surface, E is determined from the nomogram (Fig. 2.1) or from Eq. [2.11], and the correction factor ψ_E is obtained from Eq. [2.15].

2.4 CONVECTIVE HEAT-TRANSFER COEFFICIENT (h_c)

2.4.1 Definition

The convective heat-transfer coefficient depends on the velocity and physical properties of the gas flow and on geometric characteristics of the tube bundle.

2.4.2 Calculation of Design Gas Velocity

The design gas velocity is determined from the equation:

$$u_g = \frac{G_g}{F \cdot \rho_g} = \frac{G_g \cdot v_g}{F} \qquad [2.20]$$

At pressures close to atmospheric, u_g can be determined using the equation:

$$u_g = \frac{V(\vartheta + 273)}{F \cdot 273} \qquad [2.20a]$$

where V is the volumetric flow rate of gases at $P = 760$ mm Hg and $0°C$, m^3/s, and F is the minimum flow area for gas, m^2.

The minimum flow area in staggered bundles can be located in the plane of transverse spacing and in the plane of diagonal spacing alike, which is determined by the value of the parameter:

$$\varphi_{cl} = \frac{S_1 - d_{cl}}{S_1' - d_{cl}} \qquad [2.21]$$

Here, d_{cl} is the conventional diameter of the finned tube:

$$d_{cl} = d + \frac{2 \cdot l_r \cdot \delta_r}{S_r} \qquad [2.22]$$

For $\varphi_{cl} \leq 2$, the minimum free area is situated in the plane of transverse spacing and determined from the following equation:

$$F = a \cdot b - z_1 \cdot L_{c \cdot cr \cdot s} \cdot d_{cl} \qquad [2.23]$$

where a and b are the dimensions of the gas conduit in the calculated cross-section, m; $L_{c \cdot cr \cdot s}$ is the length of the tube projection on the calculated cross-section of the gas conduit, m; z_1 is the number of tubes in the transverse row.

For $\varphi_{cl} > 2$, the minimum free area lies in the plane of diagonal spacing and is calculated from the equation:

$$F = (a \cdot b - z_1 \cdot L_{c \cdot cr \cdot s} \cdot d_{cl}) \frac{2}{\varphi_{cl}} \qquad [2.24]$$

The minimum free area for in-line bundles is calculated from Eq. [2.23].

2.4.3 Physical Properties of Air and Flue Gases

Physical properties of the external heat-transfer medium, which is usually air or flue gases, are determined at the average flow temperature in the bundle ϑ.

Table 2.1 presents the kinematic viscosity ν, the thermal conductivity k, and the Prandtl number (**Pr**) for air and flue gases with the average composition at a pressure of 760 mm Hg and temperature ranging from 0 to 1000°C. The average composition of gases corresponds to the volume fractions of water vapor and carbon dioxide, which are $r_{H_2O} = 0.11$ and $r_{CO_2} = 0.13$.

The deviation of the values of ν, k, and **Pr** for flue gases with the composition different from the average one results mainly from the change in the content of water vapor. It is allowed for, correspondingly, by the corrections M_ν, M_k, and M_{Pr} to the values of ν_{av}, k_{av}, and **Pr**$_{av}$, determined from Table 2.1 for the average composition:

$$\nu = M_\nu \cdot \nu_{av} \qquad [2.25]$$

$$k = M_k \cdot k_{av} \qquad [2.26]$$

$$\mathbf{Pr} = M_{Pr} \cdot \mathbf{Pr}_{av} \qquad [2.27]$$

The corrections M_ν, M_k, and M_{Pr} are determined as functions of the gas temperature and volume fraction of water vapor from Fig. 2.2.

Table 2.1 Physical properties of air and flue gases with the average composition [11]

	Air			Flue gases of average content		
T, °C	$\nu \times 10^6$ m^2/s	$k \times 10^2$ W/m K	Pr	$\nu \times 10^6$ m^2/s	$k \times 10^2$ W/m K	Pr
0	13.6	2.42	0.70	11.9	2.27	0.74
100	23.5	3.18	0.69	20.8	3.12	0.70
200	35.3	3.89	0.69	31.6	4.00	0.67
300	48.9	4.47	0.69	43.9	4.82	0.65
400	63.8	5.03	0.70	57.8	5.68	0.64
500	73.2	5.60	0.70	73.0	6.54	0.62
600	98.0	6.14	0.71	89.4	7.40	0.61
700	116	6.65	0.71	107.0	8.25	0.60
800	136	7.12	0.72	126.0	9.13	0.59
900	157	7.59	0.72	146.0	9.99	0.58
1000	179	8.03	0.72	167.0	10.87	0.58

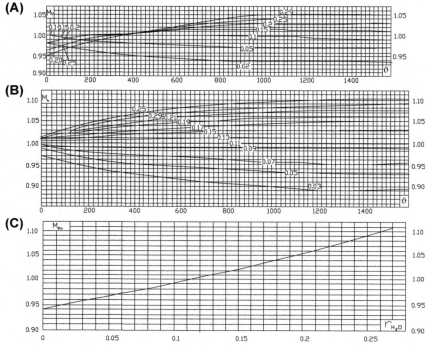

Figure 2.2 Corrections for conversion of physical characteristics of flue gases with average composition: (A) M_v; (B) M_k; and (C) M_{Pr}.

2.4.4 Calculation of Convective Coefficient in Case of Transverse Flow

The convective heat-transfer coefficient with a transverse flow around staggered and in-line bundles of tubes with circular, helical, and square finning, referred to the total surface on the gas side, is calculated from the equation:

$$h_c = 1.13 \cdot C_z \cdot C_q \cdot \frac{k_g}{d} \cdot \left(\frac{u_g \cdot d}{\nu_g}\right)^n \cdot \mathbf{Pr}_g^{0.33} \qquad [2.28]$$

where the following holds:

$$n = 0.7 + 0.08 \cdot \tanh(X) + 0.005 \cdot \Psi_r \qquad [2.29]$$

$$C_q = (1.36 - \tanh(X)) \cdot \left(\frac{1.1}{\Psi_r + 8} - 0.014\right) \qquad [2.30]$$

The shape parameter of the bundle X, entering into Eqs. [2.29] and [2.30], for the bundle with staggered arrangement of tubes is obtained from the equation:

$$X = \frac{\sigma_1}{\sigma_2} - \frac{1.26}{\Psi_r} - 2 \tag{2.31}$$

and for in-line bundles, from this equation:

$$X = 4 \cdot \left(2 + \frac{\Psi_r}{7} - \sigma_2 \right) \tag{2.32}$$

The fin coefficient for circular fins is

$$\Psi_r = \frac{1}{2d \cdot S_r} (D^2 - d^2 + 2 \cdot D \cdot \delta_2) + 1 - \frac{\delta_1}{S_r} \tag{2.33}$$

and for square fins,

$$\Psi_r = \frac{2 \left(c_{sq}^2 - 0.785 \cdot d^2 + 2c_{sq} \cdot \delta_A \right)}{\pi \cdot d \cdot S_r} + 1 - \frac{\delta_r}{S_r} \tag{2.34}$$

The factor C_z, which takes account of the effect of the number of transverse tube rows in the bundle on heat transfer z_2, for staggered bundles with $\sigma_1/\sigma_2 \geq 2$ and $z_2 < 8$, and for in-line bundles with any σ_1/σ_2 and $2 \leq z_2 < 8$ is determined from the upper curve in Fig. 2.3 or from the expression:

$$C_z = 3.5 \cdot z_2^{0.03} - 2.72 \tag{2.35}$$

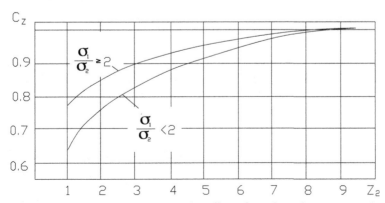

Figure 2.3 Factor C_z, taking into account the effect of number of transverse tube rows in bundle on heat transfer.

and for staggered bundles with $\sigma_1/\sigma_2 < 2$ and $z_2 < 8$, it is determined from the lower curve in Fig. 2.3 or from the expression:

$$C_z = 3.15 \cdot z_2^{0.05} - 2.50 \qquad [2.36]$$

and for $z_2 \geq 8$, as follows:

$$C_z = 1 \qquad [2.37]$$

To determine n and C_q as functions of the fin coefficient Ψ_r and the generalized arrangement parameter σ_1/σ_2 or σ_2, nomograms are constructed in Figs. 2.4 and 2.5 for staggered bundles and in Figs. 2.6 and 2.7 for in-line bundles.

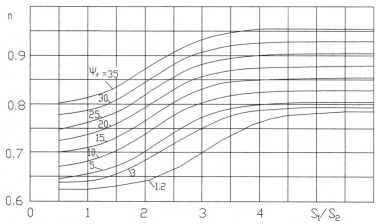

Figure 2.4 Dependence of exponent n on Reynolds number in Eq. [2.28] on geometric characteristics of staggered tube bundles.

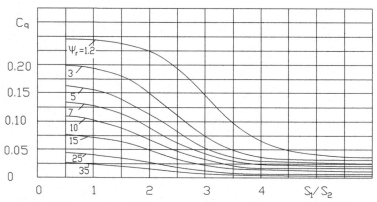

Figure 2.5 Dependence of coefficient C_q in Eq. [2.28] on geometric characteristics of staggered tube bundles.

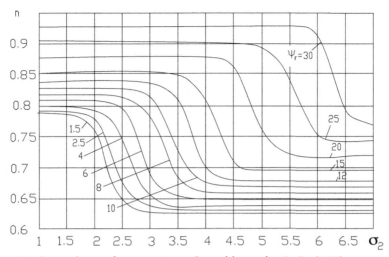

Figure 2.6 Dependence of exponent n on Reynolds number in Eq. [2.28] on geometric characteristics of in-line tube bundles.

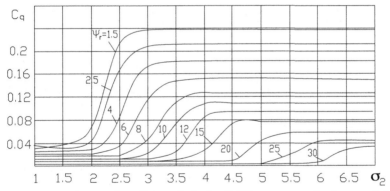

Figure 2.7 Dependence of coefficient C_q in Eq. [2.28] on geometric characteristics of in-line tube bundles.

2.4.5 Optimum Geometric Characteristics

For approximate evaluation of the heat-transfer coefficients in searching for optimum geometric characteristics of the heating surface, nomograms are constructed in Figs. 2.8– 2.17, each of which corresponds to a certain value of the Reynolds number in the range $\mathbf{Re}_d = 5 \times 10^3$ to 2.5×10^4.

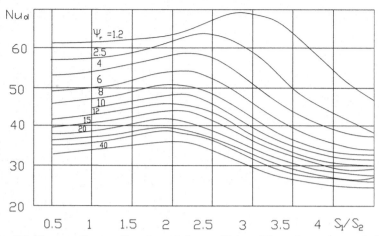

Figure 2.8 Nu_d number as function of fin coefficient Ψ_r and parameter S_1/S_2 for staggered tube bundles at $Re_d = 5000$.

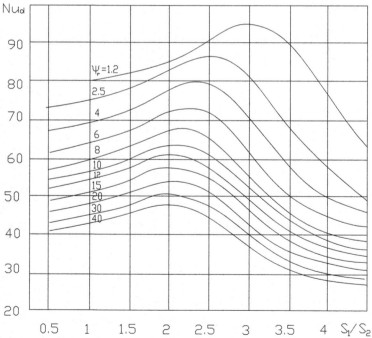

Figure 2.9 Nu_d number as function of fin coefficient Ψ_r and parameter S_1/S_2 for staggered tube bundles at $Re_d = 10,000$.

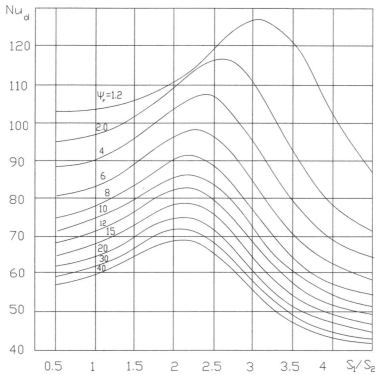

Figure 2.10 **Nu**$_d$ number as function of fin coefficient Ψ_r and parameter S_1/S_2 for staggered tube bundles at **Re**$_d$ = 15,000.

These nomograms should be used, taking into account that the Nusselt and Reynolds numbers are determined from the relations:

$$\mathbf{Nu_d} = \frac{h_c \cdot d}{k_g} \qquad [2.38]$$

$$\mathbf{Re_d} = \frac{u_g \cdot d}{\nu_g} \qquad [2.39]$$

and the fin coefficient is obtained from Eqs. [2.33] or [2.34].

It is seen from Figs. 2.8−2.17 that the dependences of the Nusselt numbers (the heat-transfer coefficients) on the arrangement parameters have a maximum, whose position is dependent mainly on the fin coefficient Ψ_r; that is, for each value of Ψ_r there is an arrangement that gives a maximum value of the heat-transfer coefficient.

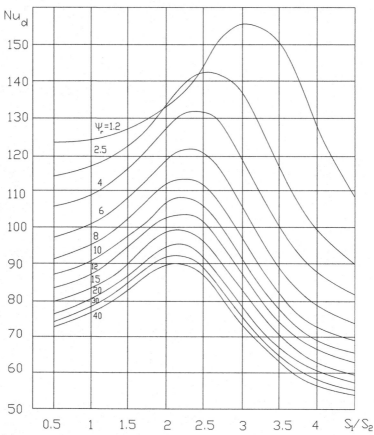

Figure 2.11 **Nu**$_d$ number as function of fin coefficient Ψ_r and parameter S_1/S_2 for staggered tube bundles at **Re**$_d$ = 20,000.

The value of the parameter $(\sigma_1/\sigma_2)_{max}$ corresponding to the maximum of the heat-transfer rate with staggered arrangement can be determined using the relation:

$$\left(\frac{\sigma_1}{\sigma_2}\right)_{max} = \frac{1.26}{\Psi_r} + 2 + \Phi \qquad [2.40]$$

and the value of the parameter (σ_2) corresponding to the maximum of the heat-transfer rate with in-line arrangement can be determined using the relation:

$$(\sigma_2)_{max} = \frac{\Psi_r}{7} + 2 - \frac{1}{4}\Phi \qquad [2.41]$$

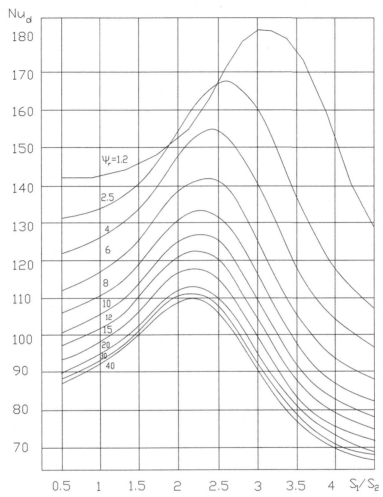

Figure 2.12 Nu$_d$ number as function of fin coefficient Ψ_r and parameter S_1/S_2 for staggered tube bundles at **Re$_d$** = 25,000.

The quantity Φ in expressions [2.40] and [2.42] is a weak function of the Reynolds number (Fig. 2.18),

$$\Phi = \frac{1}{2}\ln\left(\frac{0.189\,\ln(\mathbf{Re_d}) - 1}{1 - 0.029\,\ln(\mathbf{Re_d})}\right) \qquad [2.42]$$

At $\mathbf{Re_d} = 9800$, $\Phi = 0$. In practical calculations the quantity Φ for the range $\mathbf{Re_d} = 5000{-}25{,}000$ can be disregarded.

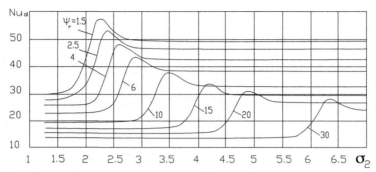

Figure 2.13 Nu$_d$ number as function of fin coefficient Ψ_r and parameter σ_2 for in-line tube bundles at **Re$_d$** = 5000.

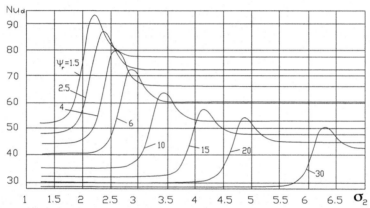

Figure 2.14 Nu$_d$ number as function of fin coefficient Ψ_r and parameter σ_2 for in-line tube bundles at **Re$_d$** = 10,000.

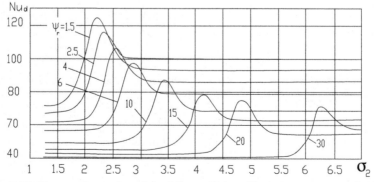

Figure 2.15 Nu$_d$ number as function of fin coefficient Ψ_r and parameter σ_2 for in-line tube bundles at **Re$_d$** = 15,000.

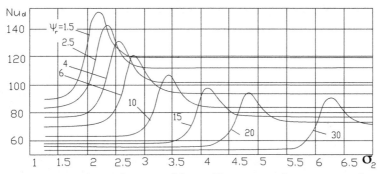

Figure 2.16 **Nu$_d$** number as function of fin coefficient Ψ_r and parameter σ_2 for in-line tube bundles at **Re$_d$** = 20,000.

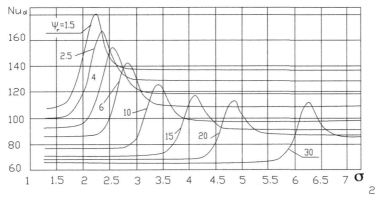

Figure 2.17 **Nu$_d$** number as function of fin coefficient Ψ_r and the parameter σ_2 for in-line tube bundles at **Re$_d$** = 25,000.

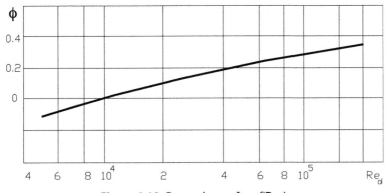

Figure 2.18 Dependence $\Phi = f(\mathbf{Re_d})$.

2.5 HEAT-TRANSFER COEFFICIENT FROM WALL TO INTERNAL MEDIUM (h_2)

2.5.1 Definition

The heat-transfer coefficient in the motion of the heat-transfer medium in tubes depends on the flow regime and velocity, physical properties of the heat-transfer medium, and geometric characteristics of tubes. In the elements of heat-transfer devices made in the form of bundles of finned tubes, there occurs, as a rule, the developed turbulent flow of the internal heat-transfer medium, which is usually water, vapor, or (in boiling) a vapor—water mixture. The thermal resistance of heat-transfer in boiling of the heat-transfer medium $1/h_2$ can be ignored. Therefore, in the calculation of the overall heat-transfer coefficient U, the quantity h_2 should be determined in the case of monophase flow in tubes.

2.5.2 Calculation in the Case of Monophase Turbulent Flow

For the case of longitudinal monophase turbulent flow over the inside surface of tubes at $\mathbf{Re}_f = 4 \times 10^3 - 5 \times 10^6$ and $\mathbf{Pr}_f = 0.1 - 2000$, the heat-transfer coefficient is determined from the equation:

$$h_2 = \frac{k_f}{d_{in}} \cdot \left[\frac{0.125 \cdot \zeta \cdot \mathbf{Re}_f \cdot \mathbf{Pr}_f \cdot C_{tem}}{\lambda + 4.5 \cdot \zeta^{0.5} \left(\mathbf{Pr}_f^{0.666} - 1 \right)} \right] \qquad [2.43]$$

where the following is calculated:

$$\lambda = 1 + \frac{900}{\mathbf{Re}_f} \qquad [2.44]$$

$$\zeta = \left(1.82 \cdot \lg(\mathbf{Re}_f) - 1.64 \right)^{-2} \qquad [2.45]$$

In narrower ranges $\mathbf{Re}_f = 1 \times 10^4 - 1 \times 10^6$ and $\mathbf{Pr}_f = 0.7 - 2.0$, the heat-transfer coefficient h_2 can be determined from the simpler equation:

$$h_2 = 0.023 \frac{k_f}{d_{in}} \mathbf{Re}_f^{0.8} \cdot \mathbf{Pr}_f^{0.4} \cdot C_{tem} \qquad [2.46]$$

In Eqs. [2.43]—[2.46], k_f is the thermal conductivity of the internal heat-transfer medium determined from its average temperature T; $\mathbf{Re}_f = \frac{u_f \cdot d_{in}}{v_f}$ is the Reynolds number determined based on the inside diameter of tubes, d_{in}.

The kinematic viscosity ν_f at the average flow temperature T ($\nu_f = \mu_f \cdot \upsilon_f$) and the medium velocity is shown:

$$u_f = \frac{G_f \cdot \upsilon_f}{f} \qquad [2.47]$$

Here, G_f is the flow rate of the internal medium, υ_f is its average specific volume (for water and water vapor is determined from Tables A.1−A.3), and f is the flow area for the passage of the heat-transfer medium, which is equal to the following:

$$f = z_{tcp} \cdot \frac{\pi \cdot d_{in}^2}{4} \qquad [2.48]$$

For coil surfaces, the number of tubes connected in parallel z_{tcp} is determined by the number of tubes in the transverse row z_1 and by the number of the coil starts n_x,

$$z_{tcp} = n_x \cdot z_1 \qquad [2.49]$$

$\mathbf{Pr_f} = \left(\frac{\mu_m c_p}{k_m}\right)$ is the Prandtl number at the average flow temperature; c_p is the specific heat at constant pressure of the medium, J/kg K; C_{tem} is the correction that takes account of the influence exerted by the temperature dependence of physical properties of the heat-transfer medium on the heat-transfer coefficient.

For dropping liquids at $\mu_w/\mu_f = 0.08-40$, the correction C_{tem} is determined from the equation:

$$C_{tem} = \left(\frac{\mu_f}{\mu_w}\right)^n \qquad [2.50]$$

Here, $n = 0.11$ for the heating of fluids; $n = 0.25$ for the cooling of fluids; μ_w is the dynamic viscosity of the medium at the average temperature of the inside surface of the tube T_w, which, with a subsequent refinement, is determined from the equation:

$$t_{in} = t + \frac{Q}{H_{in}} \cdot \frac{10^3}{\alpha_2} \qquad [2.50a]$$

μ_f is the dynamic viscosity of the medium at the average flow temperature T.

For gases, the correction C_{tem} should be determined when they are heated, using the equation:

$$C_t = \left(\frac{t + 273}{t_{\text{in}} + 273} \right)^{0.5}.$$

2.6 CONTAMINATION FACTOR AND THERMAL EFFICIENCY

2.6.1 Accounting for Contamination Factors

When combustion products of solid fuels with loose ash are used as the external heat-transfer medium, the heat-transfer coefficient is calculated with account for the contamination factors that depend on the gas velocity, the fractional composition of ash contained in the combustion products, the average temperature of gases in the bundle, geometric characteristics of the finned tubes and bundle, and on the presence of cleaning of the heating surface. The equation for the contamination factor is of the form:

$$\varepsilon = \varepsilon_0 \cdot C_d \cdot C_s \cdot C_{\text{fr}} \cdot C_\Psi \cdot C_z + \Delta\varepsilon \qquad [2.51]$$

where ε_0 is the initial contamination factor dependent on the gas velocity and the relative longitudinal spacing of tubes σ_2; C_d is the correction factor taking account of the effect of the tube diameter d; C_s is the correction factor taking account of the effect of the relative transverse spacing of tubes in the bundle σ_1; C_{fr} is the factor to account for the influence of the ash dispersity, which is determined by the residue on the screen R_{10}. When there are no reliable data on the fractional composition of ash, it is assumed that $C_{\text{fr}} = 0.9$; C_Ψ is the correction factor dependent on the fin coefficient of the tube Ψ_r and the angle of winding of fins on the tube γ, which is determined from the equation:

$$\gamma = n_{\text{wnd}} \cdot \arctan\left(\frac{s_r}{\pi \cdot d} \right) \qquad [2.52]$$

where n_{wnd} is the number of the winding starts; C_z is the correction for the number of the tube rows in the bundle in the direction of the gas motion; $\Delta\varepsilon$ is the empirical correction taking account of specific operating conditions (the type of the burned fuel, the presence of cleaning, the temperature zone, etc.).

2.6.2 Calculation of Contamination Factor in Staggered Bundles

In the calculation of staggered bundles, the initial contamination factor ε_0 and the correction factors, entering into Eq. [2.51], are determined from the graphs in Fig. 2.19. The values of the correction $\Delta\varepsilon$ are taken from Table 2.2.

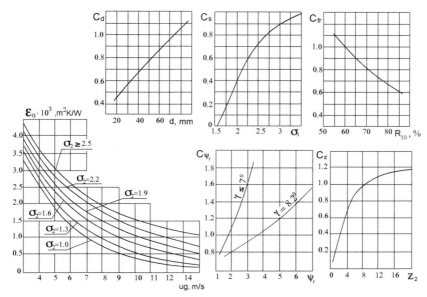

Figure 2.19 Dependences for contamination factor of staggered bundles of transversely finned tubes.

Table 2.2 Values of correction factor $\Delta\varepsilon$

| | | Correction $\Delta\varepsilon$, m^2 K/W | |
No.	Type of the heating surface	Products of combustion from solid fuels, which produce ash inlays, for cases when the surface is not cleaned	Products of combustion of other coals when there is performing periodic cleaning of surface
1.	First stages of water economizers, single-stage economizers, and other convective bundles with inlet temperature of gases $\vartheta' \leq 400°C$	0	0
2.	Second stages of water economizers, single-stage economizers with $\vartheta' > 400°C$, transition zones of once-through boilers	0.0014	–

2.6.3 Calculation of Contamination Factor in In-line Bundles

In the calculation of in-line bundles of tubes with transverse finning, the values of ε_0 and the correction factors are determined from graphs in Fig. 2.20.

The correction $\Delta\varepsilon$ for in-line tube bundles, operating in the gas temperature range $\vartheta' \leq 400°C$, is taken to be 0.0017 m^2 K/W.

2.6.4 In the Case of Combustion Products of Gas and Black Oil

When combustion products of gas and black oil are used as the external heat-transfer medium, all heating surfaces are calculated from the thermal efficiency Ψ, whose values are presented in Table 2.3.

2.6.5 In the Case of Combustion Products of a Fuel Mixture

When combustion products of a fuel mixture are used as the heat-transfer medium, contamination or thermal efficiency factors are taken for the more contaminating type of fuel.

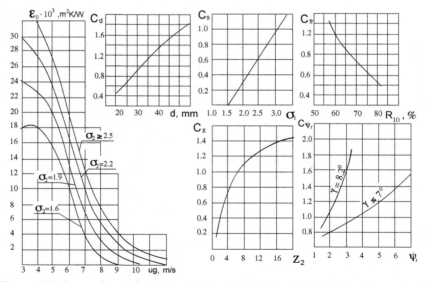

Figure 2.20 Dependences for contamination factor of in-line bundles of transversely finned tubes.

Table 2.3 Thermal efficiency factor, Ψ

Type of heating surface	Type of the burned fuel	
	Gas	Black oil
Superheaters, boiler tube banks and transition zones	0.80	0.55
"Hot" stages of economizers	0.80	0.60
"Cold" stages of economizers, single-stage economizers, preheaters of gas—liquid type	0.85	0.60
The same, with inlet temperature of water equal to 100°C and lower	0.85	0.45—0.50

2.6.6 In the Case of Flow of a Clean Heat-Transfer Medium

In the case where bundles of tubes with transverse finning are washed by the flow of a clean heat-transfer medium, for example, air (air heaters and water-to-air heat exchangers) and the passage of the entire gas flow is ensured (a complete washing), the thermal efficiency is taken to be $\Psi = 0.95$. If such surfaces are placed in air cases of complex configuration, it is necessary to assume $\Psi = 0.85$, since their washing is incomplete.

2.7 AVERAGE TEMPERATURE DIFFERENCE

2.7.1 Definition

The average temperature difference ΔT, that is, the temperature difference between the heating and the heated media averaged over the entire heating surface, depends on the relative direction of motion of the media.

2.7.2 Calculation of Temperature Difference

The connection scheme, in which both media all through the way move in parallel and in opposition, is called countercurrent. If the heat-exchanging media move in parallel in the same direction, this connection scheme is called cocurrent. The temperature difference with both schemes is determined as the logarithmic-mean temperature difference using the equation:

$$\Delta T = \frac{\Delta T_{lrg} - \Delta T_{sm}}{\ln\left(\frac{\Delta T_{lrg}}{\Delta T_{sm}}\right)} \qquad [2.53]$$

where ΔT_{lrg} is the temperature difference of the media at the end of the heating surface where it is larger, and ΔT_{sm} is the temperature difference of the media at the end of the heating surface where it is smaller.

In the cases where $\Delta T_{lrg}/\Delta T_{sm} \leq 1.7$, the temperature difference can be determined with sufficient accuracy as the arithmetic-mean temperature difference:

$$\Delta T = \frac{\Delta T_{lrg} + \Delta T_{sm}}{2} = \vartheta - T \qquad [2.54]$$

where ϑ and T are the average temperatures of the heat-exchanging media.

2.7.3 Calculation in the Case of Complex Connection Schemes

The largest possible temperature difference is achieved with a counter-current scheme, and the smallest with a cocurrent scheme. With all other connection schemes, we obtain intermediate values. Therefore, if the condition is fulfilled:

$$\Delta T_{co} \geq 0.92 \cdot \Delta T_{count} \qquad [2.55]$$

(ΔT_{co} and ΔT_{count} are the average temperature differences for the cases of cocurrent and countercurrent schemes, respectively), the temperature difference for any complex connection scheme can be determined from the equation:

$$\Delta T = 0.5(\Delta T_{co} + \Delta T_{count}) \qquad [2.56]$$

2.7.4 Calculation in the Case of Series-Mixed, Parallel-Mixed, and Crosscurrents

Given next are indications as to calculating the temperature difference for schemes different from purely cocurrent and countercurrent.

Schemes with series-mixed, parallel-mixed, and crosscurrents of the heat-exchanging media are discerned (Fig. 2.21).

The temperature differences for these schemes are determined from the equation:

$$\Delta T = \psi_{cv} \cdot \Delta T_{count} \qquad [2.57]$$

where ψ_{cv} is the factor of conversion from the countercurrent scheme to a more complex one, which is determined using appropriate nomograms.

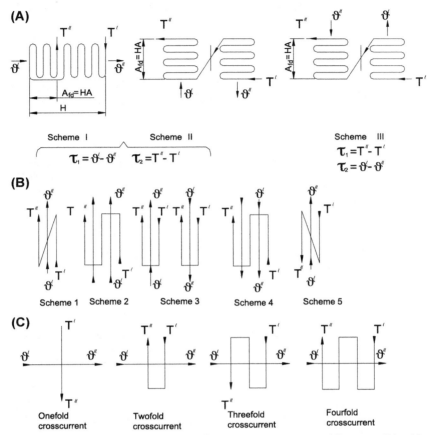

Figure 2.21 Schemes for calculation of average temperature difference: (A) with a series-mixed current; (B) with a parallel-mixed current; and (C) with a crosscurrent.

2.7.5 Dimensionless Governing Parameters for Series-Mixed Current

In the scheme with a series–mixed current, the heating surface consists of two sections connected serially with respect to both media. With transition from one section to the other, the relative motion of the media changes. For the commonest schemes of a series–mixed current, presented in Fig. 2.21A, the factor ψ_{cv} is determined from the nomogram in Fig. 2.22. Here, the following dimensionless governing parameters should be calculated:

$$H = \frac{A_{co}}{A} \qquad [2.58]$$

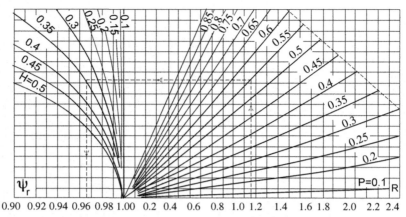

Figure 2.22 Nomogram for determining conversion factor ψ_{cv} for series-mixed current.

$$P = \frac{\tau_2}{\vartheta' - T'} \qquad [2.59]$$

$$R = \frac{\tau_1}{\tau_2} \qquad [2.60]$$

In these equations, A is the total heating surface; A_{co} is the heating surface of the cocurrent section; τ_1 and τ_2 are the total temperature drops in the heat-exchanging media:

- for schemes I and II (Fig. 2.21A), $\tau_1 = \vartheta' - \vartheta''$ and $\tau_2 = T'' - T'$;
- for scheme III, $\tau_1 = T'' - T'$ and $\tau_2 = \vartheta' - \vartheta''$

The temperature designation is given in Fig. 2.21A.

2.7.6 Dimensionless Governing Parameters for Parallel-Mixed Current

In the scheme with a parallel-mixed current (Fig. 2.21B), the heating surface consists of several sections connected serially with respect to one of the media (multipass) and in parallel with respect to the other medium (single-pass). In the calculation of the temperature difference, it is immaterial whether the heating or the heated medium is single-pass.

The factor ψ_{cv} for the schemes with a parallel-mixed current is determined from the nomogram in Fig. 2.23. The lines of its left half are used for the corresponding connection schemes.

curve 1, for schemes with two passes of a multipass medium, both of which are with a cocurrent flow relative to a single-pass medium;

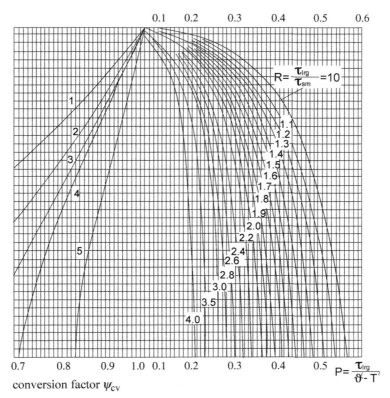

Figure 2.23 Nomogram for determining conversion factor ψ_{cv} for parallel-mixed current: 1, both passes of multipass medium are cocurrent; 2, three passes of multipass medium, of which two are cocurrent and one is countercurrent; 3, two passes of multipass medium, of which one is countercurrent and another is cocurrent; 4, three passes of multipass medium, of which two are countercurrent and one is cocurrent; and 5, both passes of multipass medium are countercurrent.

curve 2, for schemes with three passes of a multipass medium, of which two are with a cocurrent flow and one is with a countercurrent flow relative to a single-pass medium;

curve 3, for schemes with two passes of a multipass medium, of which one (no matter the first or the second) is with a countercurrent flow relative to a single-pass medium; this scheme is also used for calculating schemes with any even number of passes with the same number of countercurrent and cocurrent passes;

curve 4, for schemes with three passes of a multipass medium, of which two are with a countercurrent flow, and one is with a cocurrent flow relative to a single-pass medium;

curve 5, for schemes with two passes of a multipass medium, both of which are with a countercurrent flow relative to a single-pass medium.

The factor ψ_{cv} for schemes with an odd number of passes larger than three is taken to be equal to the half-sum of the values of ψ_{cv} determined from curves 3 and 2, or 3 and 4, depending on which passes are larger in number: cocurrent or countercurrent.

For using the nomogram in Fig. 2.23, it is necessary to calculate the dimensionless parameters:

$$P = \frac{\tau_{sm}}{\vartheta' - T'} \qquad [2.61]$$

$$R = \frac{\tau_{lrg}}{\tau_{sm}} \qquad [2.62]$$

where T' and ϑ' are the initial temperatures of the heating and the heated medium, respectively; τ_{lrg} is the total temperature drop in the medium where it is larger; and τ_{sm} is the total temperature drop in the medium where it is smaller.

2.7.7 Dimensionless Governing Parameters for Crosscurrent Scheme

In a crosscurrent scheme, the directions of flows of both media intercross. The average temperature difference for the crosscurrent depends mainly on the number of passes and the general relative direction of flows of the media (a cocurrent flow or a countercurrent flow). Crosscurrent schemes with a different number of passes are shown in Fig. 2.21C.

The factor ψ_{cv} is determined from the nomogram of Fig. 2.24. The lines of its left half are used for a different number of passes:

curve 1, for a onefold crosscurrent
curve 2, for a twofold crosscurrent
curve 3, for a threefold crosscurrent
curve 4, for a fourfold crosscurrent

For using the nomogram, the same dimensionless parameters are calculated beforehand as those for a parallel-mixed current:

$$P = \frac{\tau_{sm}}{\vartheta' - T'} \quad R = \frac{\tau_{lrg}}{\tau_{sm}}$$

2.7.8 Calculation in the Case of Steaming Economizer

If the bundle of finned tubes represents a steaming economizer with the steam quality at the exit $x \leq 30\%$, sufficient accuracy of determining the

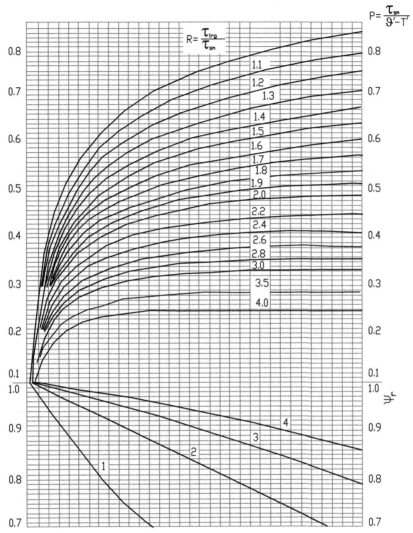

Figure 2.24 Nomogram for determining conversion factor ψ_{cv} for crosscurrent: 1, onefold crossing; 2, twofold crossing; 3, threefold crossing; and 4, fourfold crossing.

average temperature difference is provided using, as the final temperature of the heated medium, the conventional temperature:

$$T_{cl} = T_{sl} \tag{2.63}$$

where T_{cl} is the final temperature of the heated medium; T_{sl} is the saturation temperature.

CHAPTER 3

Calculation of Aerodynamic Resistance

3.1 AERODYNAMIC RESISTANCE CALCULATION

The aerodynamic resistance of the bundles of finned tubes in crossflow is determined from the Eq. [3.1]:

$$\Delta P = C_{op} \cdot \zeta_0 \cdot z_2 \cdot \frac{\rho_g \cdot u_g^2}{2}, \quad (Pa) \qquad [3.1]$$

where C_{op} is the correction factor taking account of real operating conditions of the heat-transfer surface, and it is taken to be $C_{op} = 1.1$; and ζ_0 is the resistance coefficient referred to a single transverse row of tubes.

3.1.1 Calculation of the Resistance Coefficient

The resistance coefficient ζ_0 is dependent on geometric characteristics of the bundle of finned tubes and the Reynolds number $\mathbf{Re}_{eq} = \frac{u_g \cdot d_{eq}}{\nu_g}$ and is determined in the range $\mathbf{Re}_{eq} = 5 \times 10^3$ to 6×10^4 for both staggered and in-line bundles from the equation:

$$\zeta_0 = C_z' \cdot C_r \cdot \left(\frac{u_g \cdot d_{eq}}{\nu_g} \right)^{-n} \qquad [3.2]$$

With staggered arrangement of tubes in the bundle, the exponent n and the coefficient C_r in Eq. [3.2] are defined by the expressions:

$$n = 0.17 \left(\frac{A_{total}}{F} \right)^{0.25} \left(\frac{S_1}{S_2} \right)^{0.57} \exp\left(-0.36 \frac{S_1}{S_2} \right) \qquad [3.3]$$

$$C_r = 2.8 \left(\frac{A_{total}}{F} \right)^{0.53} \left(\frac{S_1}{S_2} \right)^{1.30} \exp\left(-0.90 \frac{S_1}{S_2} \right) \qquad [3.4]$$

Nomograms are constructed (Figs. 3.1 and 3.2) for determining n and C_r using Eqs. [3.3] and [3.4].

Handbook for Transversely Finned Tube Heat Exchanger Design
ISBN 978-0-12-804397-4
http://dx.doi.org/10.1016/B978-0-12-804397-4.00003-3

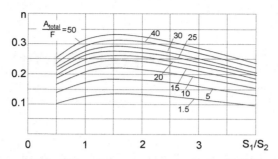

Figure 3.1 Nomogram for determining exponent n in Eq. [3.2] for calculating resistance coefficient ζ_0 of staggered bundles of finned tubes.

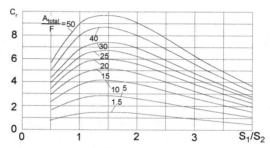

Figure 3.2 Nomogram for determining coefficient C_r in Eq. [3.2] for calculating resistance coefficient ζ_0 of staggered bundles of finned tubes.

In the case of in-line arrangement of tubes, n and C_r should be determined using the following relations:

for $\frac{S_1}{S_2} \leq 2{,}1$:

$$n = \left(\frac{A_{\text{total}}}{F}\right)^{0.08}\left(0.184 - 0.088\frac{S_1}{S_2}\right) \qquad [3.5]$$

$$C_r = 2.5\left(\frac{A_{\text{total}}}{F}\right)^{0.25}\exp\left(-1.70\frac{S_1}{S_2}\right) \qquad [3.6]$$

and for $\frac{S_1}{S_2} > 2{,}1$:

$$n = 0 \qquad [3.7]$$

$$C_r = \left(\frac{A_{\text{total}}}{F}\right)^{0.1}\left(0.132 - 0.016\frac{S_1}{S_2}\right) \qquad [3.8]$$

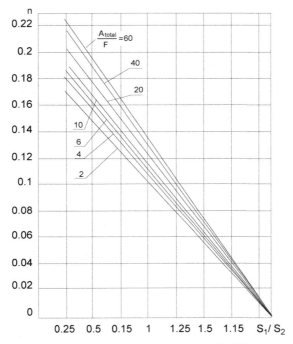

Figure 3.3 Nomogram for determining exponent n in Eq. [3.2] for calculating resistance coefficient ζ_0 of in-line bundle of finned tubes with $S_1/S_2 \leq 2.1$.

Nomograms are constructed (Figs. 3.3 and 3.4) for determining n and C_r using Eqs. [3.5]−[3.8].

The quantity $\frac{A_{total}}{F}$, entering into Eqs. [3.5]−[3.8], is called the reduced length of the developed surface and is defined by the expression:

$$\frac{A_{total}}{F} = \frac{\pi \cdot [d \cdot s_r + 2 \cdot l_r \cdot \delta_r + 2 \cdot l_r \cdot (l_r + d)]}{S_1 \cdot s_r - (d \cdot s_r + 2 \cdot l_r \cdot \delta_r)} \qquad [3.9]$$

In the calculation of the resistance coefficient ζ_0, the characteristic dimension in the Reynolds number $\mathbf{Re}_{eq}$ is taken to be the equivalent diameter of the most contracted cross-section of the bundle, which for in-line bundles and for staggered bundles with $\varphi_{cl} \leq 2$ (φ_{cl} is determined in Section 2.4.2 of Chapter 2) is shown:

$$d_{eq} = \frac{2[s_r(S_1 - d) - 2l_r\delta_r]}{2l_r + s_r} \qquad [3.10]$$

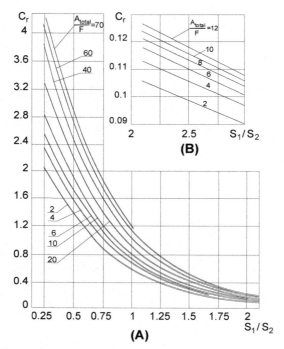

Figure 3.4 Nomogram for determining coefficient C_r in Eq. [3.2] for calculating resistance coefficient ζ_0 of in-line bundles of finned tubes: (A) $S_1/S_2 \leq 2.1$; and (B) $S_1/S_2 > 2.1$.

and for staggered bundles with $\varphi_{cl} > 2$ is as follows:

$$d'_{eq} = \frac{2d_{eq}}{\varphi_{cl}}$$ [3.11]

where d_{eq} is determined from Eq. [3.10].

For square fins, in determining $\frac{A_{total}}{F}$ from Eq. [3.9] and d_{eq} from Eq. [3.10], the following should be assumed:

$$l_A = (1.13c_{sq} - d)0.5$$ [3.12]

Depending on the arrangement of finned tubes in the bundle, the correction factor taking account of the small number of rows of the bundle C'_z is calculated using the following relations:

for staggered arrangement and $z_2 < 6$,

$$C'_z = \exp\left(0.1\left(\frac{6}{z_2} - 1\right)\right)$$ [3.13]

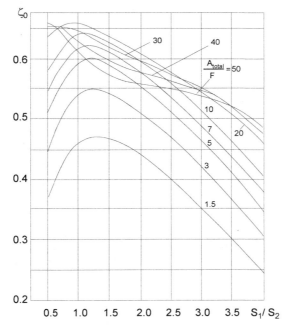

Figure 3.5 Resistance coefficient ζ_0 in Eq. [3.2] as a function of arrangement parameter S_1/S_2 and reduced length of developed surface A_{total}/F for staggered bundles of finned tubes at $Re_{eq} = 5000$.

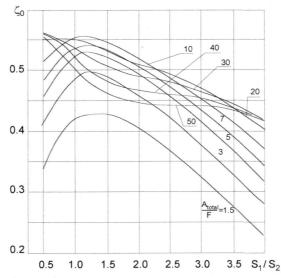

Figure 3.6 Resistance coefficient ζ_0 in Eq. [3.2] as a function of arrangement parameter S_1/S_2 and reduced length of developed surface A_{total}/F for staggered bundles of finned tubes at $Re_{eq} = 10,000$.

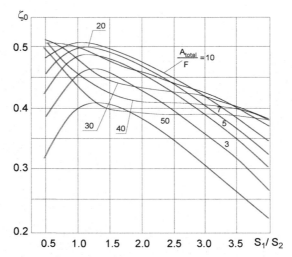

Figure 3.7 Resistance coefficient ζ_0 in Eq. [3.2] as a function of arrangement parameter S_1/S_2 and reduced length of developed surface A_{total}/F for staggered bundles of finned tubes at $\mathbf{Re_{eq}} = 15,000$.

for in-line arrangement and $z_2 < 6$,

$$C_z' = 1 + \frac{0.65}{(z_2)^3} \qquad [3.14]$$

and for any arrangement of tubes and $z_2 \geq 6$,

$$C_z' \approx 1.0 \qquad [3.15]$$

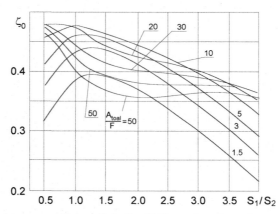

Figure 3.8 Resistance coefficient ζ_0 in Eq. [3.2] as a function of arrangement parameter S_1/S_2 and reduced length of developed surface A_{total}/F for staggered bundles of finned tubes at $\mathbf{Re_{eq}} = 20,000$.

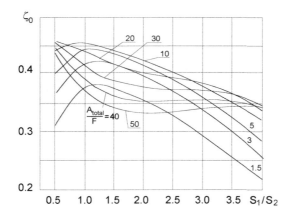

Figure 3.9 Resistance coefficient ζ_0 in Eq. [3.2] as a function of arrangement parameter S_1/S_2 and reduced length of developed surface A_{total}/F for staggered bundles of finned tubes at $\mathbf{Re}_{eq} = 25{,}000$.

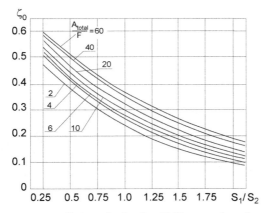

Figure 3.10 Resistance coefficient ζ_0 in Eq. [3.2] as a function of arrangement parameter S_1/S_2 and reduced length of developed surface A_{total}/F for in-line bundles of finned tubes at $\mathbf{Re}_{eq} = 5000$.

The flow velocity in Eqs. [3.1] and [3.2] is determined according to Section 2.4.2 of Chapter 2.

Physical properties of the external heat-transfer medium are determined following the recommendations given in Section 2.4.3 of Chapter 2.

Nomograms are constructed in Figs. 3.5–3.14 for approximate evaluation of the resistance coefficient ζ_0 during the search for optimum geometric characteristics of the heating surface.

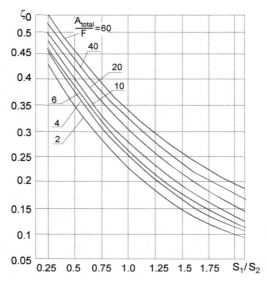

Figure 3.11 Resistance coefficient ζ_0 in Eq. [3.2] as a function of arrangement parameter S_1/S_2 and reduced length of developed surface A_{total}/F for in-line bundles of finned tubes at $\mathbf{Re_{eq}} = 10,000$.

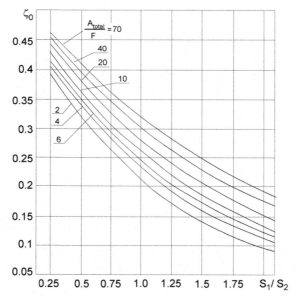

Figure 3.12 Resistance coefficient ζ_0 in Eq. [3.2] as a function of arrangement parameter S_1/S_2 and reduced length of developed surface A_{total}/F for in-line bundles of finned tubes at $\mathbf{Re_{eq}} = 15,000$.

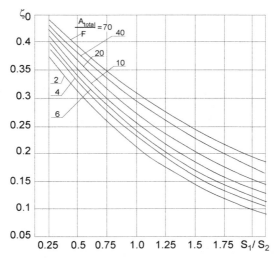

Figure 3.13 Resistance coefficient ζ_0 in Eq. [3.2] as a function of arrangement parameter S_1/S_2 and reduced length of developed surface A_{total}/F for in-line bundles of finned tubes at $\mathbf{Re}_{eq} = 20,000$.

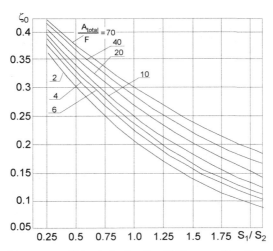

Figure 3.14 Resistance coefficient ζ_0 in Eq. [3.2] as a function of arrangement parameter S_1/S_2 and reduced length of developed surface A_{total}/F for in-line bundles of finned tubes at $\mathbf{Re}_{eq} = 25,000$.

CHAPTER 4

Calculation of Hydraulic Resistance

4.1 DEFINITION OF QUANTITIES THAT CHARACTERIZE FLOW

For the hydraulic calculation of the tube elements, use is made of the following quantities characterizing the flow of the heat-transfer medium [8–10]:

- the mass flux of the medium, kg/m^2 s,

$$(u\rho)_f = \frac{G_f}{f} \qquad [4.1]$$

- the linear velocity of the medium, m/s,

$$u_f = \frac{G_f \cdot v_f}{f} \qquad [4.2]$$

- the Reynolds number,

$$\mathrm{Re}_f = \frac{u_f \cdot d_{in}}{v_f} \qquad [4.3]$$

- and the vapor quality of the medium,

$$x = \frac{h - h'_{sl}}{h_{fg}} \qquad [4.4]$$

In Eqs. [4.1]–[4.4], G_f is the flow rate of the internal medium, kg/s; f is the area of the internal cross-section for the passage of the heat-transfer medium. In the calculation of the average velocity of the medium in tubes of the element, f is determined according to Section 2.5.2 of Chapter 2; v_f is the specific volume of the medium, m^3/kg, determined in the calculation of local resistances that are concentrated near the inlet to the tube

Handbook for Transversely Finned Tube Heat Exchanger Design
ISBN 978-0-12-804397-4
http://dx.doi.org/10.1016/B978-0-12-804397-4.00004-5

element or the outlet from it, respectively, from the temperatures of the medium at the inlet T' or outlet T''. For all other resistances, it is determined from the average temperature of the medium in the tube element average temperature; v_f is the kinematic viscosity of the medium determined from its average temperature; i_{fg} is the latent heat of evaporation, kJ/kg; h'_{sl} is the enthalpy of the liquid phase of the internal medium on the saturation line, kJ/kg; h is the enthalpy of the medium in the calculated cross-section or the average enthalpy in the segment (depending on which x should be determined—that in the calculation cross-section or that average in the segment), kJ/kg.

4.2 CALCULATION OF TOTAL HYDRAULIC RESISTANCE OF TUBE ELEMENT

The total hydraulic resistance of the tube element is expressed by the Eq. [9]:

$$\Delta P_{el} = \Delta P_{fr} + \sum \Delta P_{loc} + \overline{\Delta P_{col}} \qquad [4.5]$$

where ΔP_{fr} is the hydraulic friction resistance, Pa, which is determined according to Section 4.4; $\Sigma \Delta P_{loc}$ is the pressure losses at local resistances, Pa, which are determined according to Section 4.5; and $\overline{\Delta P_{col}}$ is the overall pressure loss in the element collectors referred to the tube with the mean flow rate of the medium, Pa, which is determined according to Section 4.6.4.

4.3 CALCULATION OF HYDRAULIC FRICTION AND LOCAL RESISTANCES OF HEAT-TRANSFER SURFACE

The hydraulic friction resistance and local resistances of the heat-transfer surfaces, in which a phase transition occurs, for example, of evaporation surfaces and steaming economizers, are calculated separately for the segments with monophase and two-phase flows, the economizer (L_{ec}) and the evaporation (L_{ev}) segments, respectively. The length of each of these segments is determined from the following equations:

$$L_{ec} = L_{un} + L_h \frac{h'_{sl} - h'}{\Delta h_{el}} \qquad [4.6]$$

$$L_{ev} = L_{total} - L_{ec} \qquad [4.7]$$

where L_{total} is the total length of tubes (coils) of the element from the supply collector to the intake collector, m; L_{un} is the length of unheated tube segments, m; $L_h = L_{rs} + L_t$ is the average length of the heated tube segments, m; h' is the enthalpy of the medium at the entrance to the element, kJ/kg; h'_{sl} is the enthalpy of the liquid phase on the saturation line, kJ/kg; and $\overline{\Delta h_{el}}$ is the average increment of the enthalpy of the medium in the considered element, kJ/kg.

4.4 FRICTION RESISTANCE

4.4.1 Calculation of Friction Resistance of Monophase Liquid Motion

The friction resistance in the motion of a monophase liquid in tubes is calculated using Eq. [4.8]:

$$\Delta P_{fr} = \zeta_{fr} \frac{L}{d_{in}} \frac{\rho_f u_f^2}{2} \qquad [4.8]$$

where ζ_{fr} is the coefficient of hydraulic friction resistance, which is determined according to Section 4.4.2; L is the length of the calculated tube segment that is taken to be $L = L_{total}$ in the calculation of surfaces with a monophase flow and $L = L_{ec}$ in the calculation of surfaces in which a phase transition occurs; ρ_f is the density of the medium, kg/m^3, determined from the average temperature and the average pressure of the medium; u_f is the linear velocity of the medium, m/s, which is determined according to Section 4.1 or Section 2.5.2 of Chapter 2.

4.4.2 Determination of Absolute Tube Roughness

The coefficient of hydraulic friction resistance ζ_{fr} is a function of the Reynolds number Re_f (Section 4.1) and the relative roughness of tubes $\ni / d_{in}$.

The absolute roughness of tubes $\ni$ for various materials of which tubes are made is determined according to Table 4.1.

Table 4.1 Absolute tube roughness, $\ni$

Material	$\ni$ (m)
Carbon and alloyed (perlite) steels	8.0×10^{-5}
Austenitic steels	1.0×10^{-5}
Aluminum	1.5×10^{-5}

The most characteristic feature of engineering heat-transfer devices, made in the form of coil tube elements, is the developed turbulent flow of the internal medium. Therefore, the values of the friction factor ζ_{fr} for $Re_f \geq 560 \cdot d_{in}/\ni$ (in the self-similar region) can be calculated from the equation:

$$\zeta_{fr} = \frac{1}{4\left(\lg\left(3.7\frac{d_{in}}{\ni}\right)\right)^2} \qquad [4.9]$$

In the cases where the self-similar flow is not reached (for example, when water moves in small-diameter tubes with $T_f < 150°C$ and $u_f \leq 0.3$ m/s), it is expedient to determine the friction factor from Fig. 4.1.

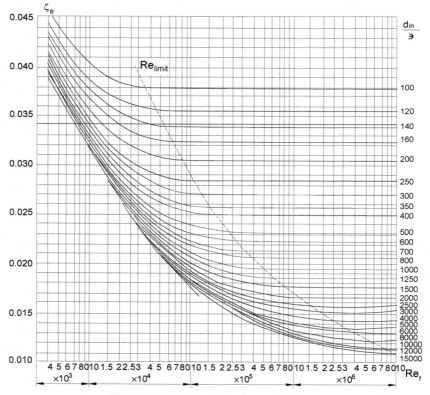

Figure 4.1 Friction factor in rough tubes.

4.4.3 Calculation of Friction Resistance of Two-Phase Liquid Motion

When a two-phase liquid moves in tubes and the vapor quality of the vapor—liquid mixture varies along the length of the segment, the friction resistance is obtained from the equation:

$$\Delta P_{fr} = \zeta_{fr} \frac{L_{ev}}{d_{in}} \frac{(u\rho)_f^2}{2\rho'_{sl}} \left[1 + \overline{\psi}_{tw \cdot ph} \overline{x} \left(\frac{\rho'_{sl}}{\rho''_{sl}} - 1 \right) \right] \qquad [4.10]$$

where ζ_{fr} is the friction factor of a monophase flow, determined according to Section 4.4.2 for a flow of the same amount of liquid; $(u\rho)_f$ is the mass velocity of liquid, kg/m²s, determined according to Section 4.1; ρ'_{sl} is the liquid density on the saturation line, kg/m³; ρ''_{sl} is the dry vapor density on the saturation line, kg/m³; $\overline{x}$ is the average vapor quality in the evaporation segment, determined according to Section 4.1 using Eq. [4.4]. In the case where the considered section coincides with that where evaporation begins, $\overline{x}$ can be determined using the equation:

$$\overline{x} = \frac{h'' - h'_{sl}}{2h_{fg}} \qquad [4.11]$$

Here, h'' is the enthalpy of the medium at the outlet from the element; $\psi_{tw \cdot ph}$ is the coefficient, determined from Eq. [4.12] and Fig. 4.2,

$$\overline{\psi}_{tw \cdot ph} = \frac{\psi''_{tw \cdot ph} x'' - \psi'_{te \cdot ph} x'}{x'' - x'} \qquad [4.12]$$

where x' and x'' are the vapor qualities of the medium at the inlet to the considered segment and at the outlet from it, which are determined according to Section 4.1; $\psi'_{tw \cdot ph}$ and $\psi''_{tw \cdot ph}$ are the quantities determined according to Fig. 4.2 from the initial x' and final x'' vapor qualities in the considered segment.

4.5 LOCAL RESISTANCE IN TUBE ELEMENTS

The local resistances in the tube elements are, as a rule, composed of local resistances of the entrance to tubes from collectors and of the exit from tubes to collectors, and of local resistances of the flow turns (the tube bends).

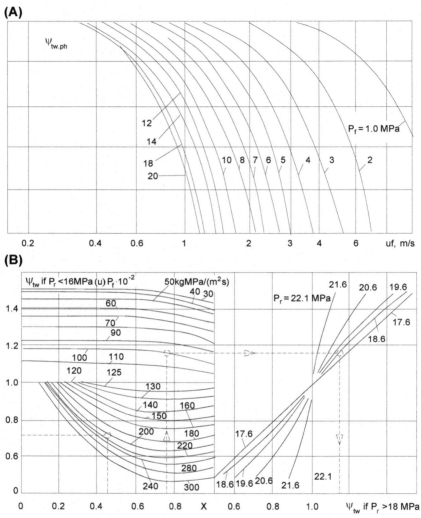

Figure 4.2 Values of coefficient $\psi_{tw\cdot ph}$ in Eq. [4.12]: (A) for $x < 0.7$ and $u_f < 10$ m/s; and (B) for $u_f > 10$ m/s.

4.5.1 Calculation of Local Resistance in Monophase Liquid Motion

The local resistances in the motion of a monophase liquid in tubes are determined from the equation:

$$\Delta P_{\text{loc}} = \zeta_{\text{loc}} \frac{\rho_f u_f^2}{2} \qquad [4.13]$$

where ζ_{loc} is the local resistance coefficient, determined according to the type of the local resistance using the data in Sections 4.5.2–4.5.4; ρ_f is the density of the medium, kg/m^3, determined according to Section 4.1 in the same manner as is the specific volume v_f; u_f is the flow velocity, m/s, to which we refer the local resistances.

4.5.2 Determining Average Resistance Coefficients

The average resistance coefficients of the entrance to the tube ζ_{ent} are determined from Table 4.2 or Fig. 4.3.

Table 4.2 Average resistance coefficients of tube entrance, ζ_{ent}

Type of tube entrance	Resistance coefficient	
	$\frac{d_{in}}{d_{col}} \leq 0.1$	$\frac{d_{in}}{d_{col}} > 0.1$
To heated tube from distributing collector with end or side supply of medium (Fig. 4.3, case 2)	0.5	0.7
To heated tube from distributing collector with radial supply of medium and with number of outlet tubes $n \leq 30$ per one inlet tube (Fig. 4.3, case 1)	0.5	0.7
To heated tube from distributing collector with radial supply of medium and with number of outlet tubes $n > 30$ per one inlet tube (Fig. 4.3, case 1)	0.6	0.8
To outlet tube of intake collector with end or side outlet of medium (Fig. 4.3, case 3)	0.4	0.4
To outlet tube of intake collector with distributed radial outlets of medium (in active zone) (Fig. 4.3, case 4)	0.5	0.5

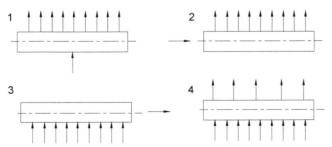

Figure 4.3 Diagrams for calculation of resistance coefficients of tube entrance and exit: 1, distribution collector with radial inlet; 2, distribution collector with side inlet; 3, intake collector with end outlet; and 4, intake collector with distributed radial outlets in active zone.

Table 4.3 Average resistance coefficients of tube exit, ζ_{ex}

Type of tube exit	ζ_{ex}
To collector with end inlet of medium	0.8
To collector with radial inlet of medium in active zone	1.1
To collector with side inlet of medium	1.3

4.5.3 Determining Resistance Coefficient of Tube Exit

The average resistance coefficients of the exit from the tube ζ_{ex}, related to the velocity in it, are determined from Table 4.3.

4.5.4 Determining Resistance Coefficient of Tube Turns

The resistance coefficients of tube turns (bends) ζ_{bnd} are assumed using Fig. 4.4 or Table 4.4 depending on the relative radius of the bend R/d_{in} and the angle of turn φ.

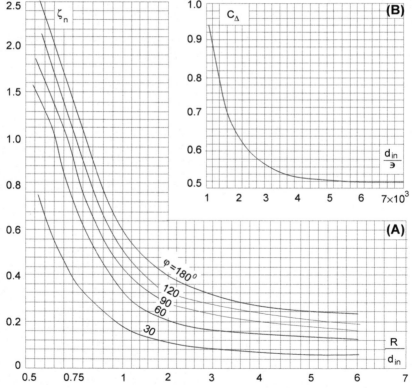

Figure 4.4 Nomograms for determining resistance coefficients of bends ($\zeta_{bnd} = \zeta_n \cdot C_\Delta$); C_Δ is correction coefficient on roughnesses that are smaller than those adopted on the main area of nomogram: (A) $\frac{d_{int}}{\xi} \leq 1250$; and (B) $\frac{d_{int}}{\xi} > 1250$.

Table 4.4 Resistance coefficients of turns (bends), ζ_{bnd}

Angle of turn φ	Radius of the bend $\frac{R}{d_{in}}$					
	1.0	1.5	2.0	3.0	4.0	5.0
60°	0.32	0.25	0.21	0.18	0.16	0.15
90°	0.41	0.34	0.3	0.24	0.22	0.2
120°	0.5	0.4	0.34	0.27	0.24	0.22
180°	0.6	0.48	0.4	0.32	0.28	0.26

4.5.5 Calculating Resistance in Motion of Two-Phase Liquid

The local resistances in the motion of a two-phase liquid in tubes are determined from the equation:

$$\Delta P_{loc} = \sum \zeta'_{loc} \frac{(u\rho)_f^2}{2\rho'_{sl}}\left[1 + \bar{x}\left(\frac{\rho'_{sl}}{\rho''_{sl}} - 1\right)\right]$$ [4.14]

where $\sum \zeta'_{loc}$ is the sum of conventional local resistance coefficients, determined according to Sections 4.5.6—4.5.8. The other quantities in Eq. [4.14] are determined in the same manner as in Section 4.4.3.

4.5.6 Determining Coefficients of Two-Phase Flow Entrance

The conventional resistance coefficients of the entrance of a two-phase flow ζ'_{ent} to vertical and inclined tubes from collectors are assumed using Table 4.5 depending on the relative tube height L_h/d_{in}.

The conventional resistance coefficients of the mixture entrance to horizontal tubes are taken to be equal to the resistance coefficients for a uniphase medium (Section 4.5.2).

4.5.7 Resistance Coefficient of Exit of a Two-Phase Medium

The conventional resistance coefficients of the exit of a two-phase medium from the tube to the enclosure ζ'_{ex}, including to collectors with distributed supply and withdrawal, are taken to be the following:

$$\zeta'_{ex} = 1.2$$

Table 4.5 Resistance coefficients of two-phase flow entrance, ζ'_{ent}

Arrangement of tubes	Relative tube height $\frac{L_h}{d_{in}}$							
	10	20	50	≥80	10	20	50	≥80
	$P \leq 6.0$ MPa				$P > 6.0$ MPa			
Vertical	0.3	0.5	0.8	1.0	0.6	0.9	1.1	1.2
At an angle, with progression to a vertical	0.5	1.1	1.7	2.2	1.0	1.2	1.4	1.5

4.5.8 Determining Resistance Coefficient of Turns in Two-Phase Flow

The resistance coefficients of turns (bends) ζ'_{bnd} in a two-phase flow depend on the arrangement and length of the tube segments behind the turn L_φ.

For the mass velocities of the medium $(u\rho)_f \leq 1200 \text{ kg/m}^2 \text{ s}$, on the average, the following values of ζ'_{bnd} are assumed:

A horizontal or small $(L_\varphi/d_{in} \leq 10)$ vertical or inclined segment (for example, with a turn ahead of the entrance to the collector)	$\zeta'_{bnd} = \zeta_{bnd}$
An inclined segment $(L_\varphi/d_{in} > 10)$ with the angle of elevation not larger than 15 degree	$\zeta'_{bnd} = 2\zeta_{bnd}$
A vertical or inclined segment with the angle of elevation larger than 15 degree and $L_\varphi/d_{in} > 10$	$\zeta'_{bnd} = 4\zeta_{bnd}$
A lowering, vertical, or inclined segment (the angle of turn is larger than 90 degree) with $L_\varphi/d_{in} > 10$	$\zeta'_{bnd} = 2\zeta_{bnd}$

ζ_{bnd} is the resistance coefficient of a given turn with the motion of a monophase medium (Section 4.5.4).

With the mass velocities $(u\rho)_f > 1200 \text{ kg/m}^2 \text{ s}$ for all schemes, the values of the resistance coefficients of turns ζ'_{bnd} are taken as for a monophase medium (Section 4.5.4),

$$\zeta'_{bnd} = \zeta_{bnd}.$$

4.6 PRESSURE VARIATION IN COLLECTORS [9]

4.6.1 Definition

In convective heat-transfer surfaces, horizontal collectors are generally used that supply the medium through tubes (distributing collector (DC) or supply collector (SC)) and collect it from tubes (intake collector (IC)). A part of the collector, into which the medium is fed or from which it is withdrawn, is called the active part.

According to the arrangement of inlet and outlet tubes, collectors with "radial" and "end" supply or withdrawal of the medium are distinguished. The radial supply (withdrawal) can be carried out both within and beyond the active zone. In the first case, the inlets (the outlets) are placed uniformly lengthwise, and such a collector is called a collector with uniform supply or withdrawal. A nonuniform positioning of the radial inlets or outlets is also possible. The radial supply (withdrawal) of the medium beyond the active zone is called lateral or angular.

4.6.2 Accounting for Static Pressure

The static pressure varies along the length of collectors as a result of the change in the medium velocities and of nonrecoverable losses due to friction and at local resistances. The variation in the static pressure along the length of the collector influences the pressure difference between the inlet and outlet cross-sections of the attached tubes, causing a nonuniform distribution of the medium between them (a hydraulic maldistribution). This should be taken into account especially with a unidirectional supply or withdrawal of the medium. The variation in static pressure along the length of collectors can be disregarded when the medium is uniformly supplied to or withdrawn from them by tubes, located in no less than three cross-sections of its active part, and when the cross-sectional area of the collector is larger than the total cross-sectional area of all attached inlet (outlet) tubes. For the evaporation and economizer surfaces, the pressure drop in collectors may be not checked when only the second condition is fulfilled.

The total hydraulic resistance of the tube element is determined taking into account the static pressure losses in the collector segments from the entrance of the medium (the exit for IC) to the tube with the mean flow rate of the medium.

4.6.3 Calculation of Variation in Static Pressure

The maximum variation in the static pressure along the length of a horizontally positioned collector is determined from the equation:

$$\Delta P_{col} = B \frac{\rho_f u_{fmax}^2}{2}, \qquad [4.15]$$

where u_{fmax} is the maximum velocity of the medium in the collector (at the entrance to SC and at the exit from IC); ρ_f is the medium density in the collector; B is the coefficient taking into account losses in the collector, which is assumed according to Table 4.6.

The values of B, indicated in Table 4.6 for a radial supply of the medium, refer to inlet tubes whose axes are at angles between 60 and 120 degrees to the axes of heated tubes. With other angles (especially with those close to 180°), the values of the coefficient B can be markedly different from those given in Table 4.6. The hydraulic maldistribution with angles larger than 120 degree can noticeably increase; therefore inlet tubes should not be positioned at such angles to heated tubes.

Table 4.6 Values of coefficient B

Type of the collector	B
For intake collectors (ICs):	
With radial outlet in center of active part	1.8
With end outlet	2.0
For distributing collectors (DCs):	
With radial inlet in center of active part and $f_{col}/f_{inl} = 1.0$	1.6
With radial inlet in center of active part and $f_{col}/f_{inl} = 1.5$	2.0
With end inlet with full cross-section and $f_{col} = f_{inl}$	0.8
With end inlet with not full cross-section and $f_{col} > f_{inl}$	$2\left(\frac{f_{col}}{f_{inl}} - 0.6\right)$
With angular inlet (beyond active zone)	1.0

In Table 4.6, f_{col} is the area of the internal cross-section of the collector, and f_{inl} is the overall cross-sectional area of all tubes feeding the medium to the supply collector.

4.6.4 Calculation of Static Pressure Loss

The total static pressure loss in the horizontal supply and intake collectors of hydraulic schemes based on the tube with the mean flow rate of the medium ΔP_{col} is determined according to the scheme of the medium motion in the calculated element (Fig. 4.5) from the following equations:

- for П, Z, ПZ, ГР, ПР, ZП, and ТР schemes of the medium motion in the tube elements,

$$\overline{\Delta P}_{col} = \frac{2}{3}\left(\Delta P_{col}^{int} - \Delta P_{col}^{sup}\right) \qquad [4.16]$$

- for a unidirectional end supply and a uniform radial withdrawal ($n \geq 3$),

$$\overline{\Delta P}_{col} = -\frac{2}{3}\Delta P_{col}^{sup} \qquad [4.17]$$

- for a distributed radial supply ($n \geq 3$) and a unidirectional end withdrawal,

$$\overline{\Delta P}_{col} = \frac{2}{3}\Delta P_{col}^{int} \qquad [4.18]$$

- for distributed coaxial supply ($n \geq 3$) and withdrawal ($n \geq 3$),

$$\overline{\Delta P}_{col} = 0; \qquad [4.19]$$

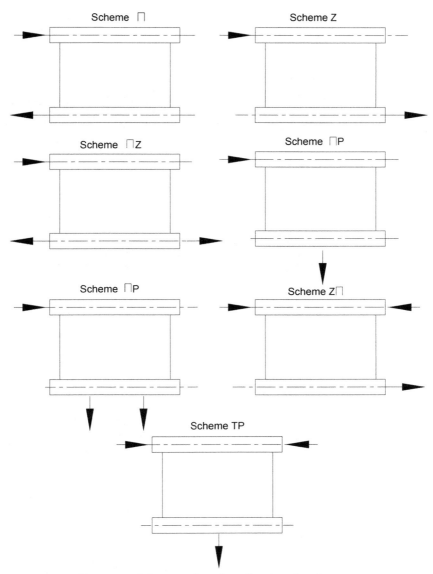

Figure 4.5 Schemes of medium flow in tube elements.

- for a concentrated supply in the center of the active part of the collector and a uniformly distributed withdrawal ($n \geq 3$),

$$\overline{\Delta P}_{\mathrm{col}} = -\frac{2}{3}\Delta P_{\mathrm{col}}^{\mathrm{sup}}$$ [4.20]

In Eqs. [4.16]–[4.20], $\Delta P_{\mathrm{col}}^{\mathrm{sup}}$ and $\Delta P_{\mathrm{col}}^{\mathrm{int}}$ are the maximum variations in the static pressure in the supply and intake collectors of hydraulic schemes, which are determined according to Section 4.6.3.

4.6.5 Calculation of Pressure Drops

The difference of total pressure drops in the supply and intake collectors of the calculated element for cross-sections with a maldistributed tube, and the mean flow rate of the medium δP_{col} is determined according to the scheme of the medium motion in the element (Fig. 4.5) and the value of the ratio $\Delta P_{\mathrm{col}}^{\mathrm{sup}}/\Delta P_{\mathrm{col}}^{\mathrm{int}}$ using the following equations:

- for the Π scheme,

$$\delta P_{\mathrm{col}} = \frac{1}{3}\left(\Delta P_{\mathrm{col}}^{\mathrm{int}} - \Delta P_{\mathrm{col}}^{\mathrm{sup}}\right) \quad \text{for} \quad \frac{\Delta P_{\mathrm{col}}^{\mathrm{sup}}}{\Delta P_{\mathrm{int}}^{\mathrm{ins}}} < 1;$$ [4.21]

$$\delta P_{\mathrm{col}} = \frac{2}{3}\left(\Delta P_{\mathrm{col}}^{\mathrm{sup}} - \Delta P_{\mathrm{int}}^{\mathrm{ins}}\right) \quad \text{for} \quad \frac{\Delta P_{\mathrm{col}}^{\mathrm{sup}}}{\Delta P_{\mathrm{int}}^{\mathrm{ins}}} > 1;$$ [4.22]

- for the ΠZ scheme,

$$\delta P_{\mathrm{col}} = \frac{\left(2\Delta P_{\mathrm{col}}^{\mathrm{int}} - \Delta P_{\mathrm{col}}^{\mathrm{sup}}\right)^2}{3\left(4\Delta P_{\mathrm{col}}^{\mathrm{int}} - \Delta P_{\mathrm{col}}^{\mathrm{sup}}\right)} \quad \text{for} \quad \frac{\Delta P_{\mathrm{col}}^{\mathrm{sup}}}{\Delta P_{\mathrm{col}}^{\mathrm{int}}} < 4;$$ [4.23]

$$\delta P_{\mathrm{col}} = \frac{2}{3}\Delta P_{\mathrm{col}}^{\mathrm{sup}} - \frac{2}{3}\Delta P_{\mathrm{col}}^{\mathrm{int}} \quad \text{for} \quad \frac{\Delta P_{\mathrm{col}}^{\mathrm{sup}}}{\Delta P_{\mathrm{col}}^{\mathrm{int}}} \geq 4;$$ [4.24]

- and for the Z, ΓP, ΠP, $Z\Pi$, and TP schemes,

$$\delta P_{\mathrm{col}} = \frac{2}{3}\Delta P_{\mathrm{col}}^{\mathrm{sup}} + \frac{1}{3}\Delta P_{\mathrm{int}}^{\mathrm{ins}}$$ [4.25]

In Eqs. [4.21]–[4.25], $\Delta P_{\mathrm{col}}^{\mathrm{sup}}$ and $\Delta P_{\mathrm{col}}^{\mathrm{int}}$ are the maximum variations in the static pressure in the supply and intake collectors of hydraulic schemes, which are determined according to Section 4.6.3.

CHAPTER 5

Calculation of Temperature Mode of Finned Tubes

5.1 DEFINITION

To evaluate the reliability and to perform the strength design for transversely finned tubes, operating under the conditions of high heat loads and high temperatures of the heat-exchanging media, the calculation of their temperature mode is done by determining [13] the following:

- the temperatures of the fin base and tip
- the mean integral temperature of the fin
- the temperature of the inside surface of the tube

The temperature mode is calculated at the places of the heat-transfer surfaces, where the specific heat absorption and the temperature of the heated medium are high, and the hydraulic maldistribution is large. The combination of these factors can also result in large values of these temperatures (Fig. 5.1).

5.2 CALCULATION OF TEMPERATURE OF FIN BASE

The temperature of the fin base is determined from the expression:

$$T_1 = T_{c \cdot cr \cdot s} + \Delta T_t + \beta_d \cdot \lambda \cdot q_{max} \times 10^3 \left(\Theta_k \cdot \frac{\delta_t}{k_t} \cdot \frac{2}{1 + \beta_d} + \frac{1}{h_2} \right) \quad [5.1]$$

where $T_{c \cdot cr \cdot s}$ is the average temperature of the internal (heated) medium in the calculated cross-section of the considered surface element, determined according to Section 5.2.1; ΔT_t is the excess of the temperature of the internal medium in the maldistributed tube over the average temperature in the calculated cross-section, determined according to Section 5.2.3; $\beta_d = d/d_{in}$ is the ratio of the outside and inside diameters of the finning-carrying tube; λ is the coefficient of heat distribution, taken to be $\lambda = 1$; δ_t is the wall thickness of the finning-carrying tube; k_t is the thermal conductivity of the metal of the wall of the finning-carrying tube dependent on

Handbook for Transversely Finned Tube Heat Exchanger Design
ISBN 978-0-12-804397-4
http://dx.doi.org/10.1016/B978-0-12-804397-4.00005-7

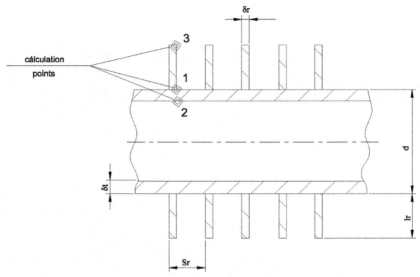

Figure 5.1 Schematic of disposition of calculation points in determining characteristic temperatures of a finned tube.

the material of the wall and its average temperature T_w; for determining k_t it is possible to assume, with a subsequent refinement:

$$T_w \approx T_{c \cdot cr \cdot s};$$ [5.2]

h_2 is the coefficient of heat transfer from the wall to the internal medium, determined according to Section 5.2.5; Θ_k is the ratio of the temperature difference between the outside and the inside surfaces of the finning-carrying tube in the cross-section, passing through the middle of the fin, to the tube-length-average value of this difference; it is calculated from the equation:

$$\Theta_k = 1 + 0.5 \cdot \frac{S_r - \delta_r}{\delta_t}$$ [5.3]

q_{max} is the heat flux at the point of the maximum heat absorption of the most loaded tube, kW/m², determined according to Section 5.2.4.

5.2.1 Calculation of Temperature of Internal Medium

The average temperature of the internal medium in the calculated cross-section of the considered surface element $T_{c \cdot cr \cdot s}$ for evaporation heat-transfer surfaces is taken to be equal to the saturation temperature. For other types of surfaces, it is determined from the medium enthalpy in the calculated cross-section:

$$h_{c \cdot cr \cdot s} = h' + \Delta h_x, \qquad [5.4]$$

where h' is the medium enthalpy at the entrance to the element, kJ/kg; Δh_x is the average increment of the medium enthalpy in the tube element up to the calculated cross-section, evaluated according to Section 5.2.2.

In this case, by the element is meant a tubular heating surface situated between two collectors (Fig. 5.2). If the heat-transfer surface does not have intermediate mixing collectors, the calculated element is the entire heat-transfer surface considered.

5.2.2 Calculation of Increment of Medium Enthalpy

The average increment of the medium enthalpy from the beginning of the element up to the calculated cross-section is as follows:

$$\Delta h_x = \eta_{wd} \cdot \frac{Q_{seg}}{G_{el}}, \text{kJ/kg} \qquad [5.5]$$

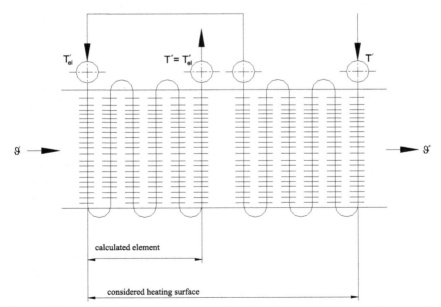

Figure 5.2 Sample of two-stage convective heat-transfer surface with complex countercurrent–co-current scheme of medium flow and intermediate mixing collector.

where η_{wd} is the coefficient of nonuniformity of the heat-transfer rate of the element across the width of the gas conduit; for the gas temperature at the entrance to the convection bundle $\vartheta' \leq 900°C$, it is possible to assume $\eta_{wd} = 1.0$ irrespective of the element position across the width of the gas conduit; G_{el} is the flow rate of the internal medium in the element, kg/s; Q_{seg} is the heat-transfer rate of the element segment up to the calculated cross-section, kW, which is determined disregarding the intertube radiation because of the small thickness of the radiative layer in the bundles of finned tube:

$$Q_{seg} = Q_{rad} + Q_c \qquad [5.6]$$

Here, Q_{rad} is the heat-transfer rate of the calculated segment by radiation from the gas volume in front of the bundle, kW, determined according to Section 5.2.2.1; and Q_c is the heat-transfer rate of the calculated segment by convection, kW, determined according to Section 5.2.2.2.

5.2.2.1 Calculation of Heat Transfer Rate by Radiation

The heat transfer rate of the calculated segment by radiation from the gas volume in front of the bundle is shown:

$$Q_r = q_{rad} A_{rad} \qquad [5.7]$$

Here, A_{rad} is the radiation-absorbing surface of the calculated segment:

$$A_{rad} = \sum_{i=1}^{n} (F_{iar} \cdot X_{irow}) \qquad [5.8]$$

F_{iar} is the area of the surface, passing through the axes of tubes of the i-th row of the bundles; x_{irow} is the angular coefficient of the i-th row of the bundle, determined from Fig. 5.3 for the first row and from Fig. 5.4 for the other tube rows; $n = z_{2seg}$ is the number of the tube rows in the direction of the gas motion within the calculated segment; q_{rad} is the specific heat absorption of the calculated segment by radiation from the preceding gas volume:

$$q_{rad} = 5.7 \times 10^{-11} a \left(T^4 - T^4_{s \cdot cont}\right) \frac{a_{cont} + 1}{2}; kW/m^2 \qquad [5.9]$$

In Eq. [5.9], T is the gas temperature in the preceding gas volume:

$$T = \vartheta' + 273; \qquad [5.10]$$

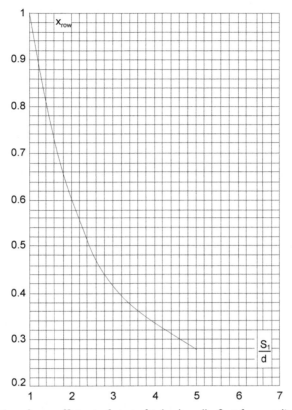

Figure 5.3 Angular coefficient of row of tube bundle first from radiation source.

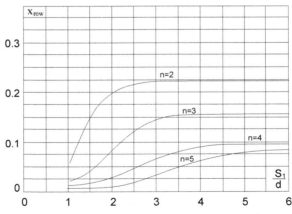

Figure 5.4 Angular coefficients of row of tube bundle second from radiation source and of subsequent rows.

a_{cont} is the emissivity factor of the contaminated tube walls, taken to be $a_{cont} = 0.8$; a is the emissivity factor of gases in the preceding gas volume, defined by the expression:

$$a = 1 - \exp(-K \cdot P \cdot s), \qquad [5.11]$$

where $P = P_g$ is the pressure in the gas conduit, MPa; s is the effective thickness of the radiative layer of the gas volume, bounded on all sides, m, which is determined from the equation:

$$s = 3.6 \frac{V_{b \cdot sp}}{F_{b \cdot sp}}, \qquad [5.12]$$

in which $V_{b \cdot sp}$ is the volume of the space in front of the bundle, m^3; $F_{b \cdot sp}$ is the area of surfaces, bounding the preceding volume, m^2; K is the radiant absorptance of the gas medium, 1/m MPa; in the case of a dust-laden gas flow over the bundle of finned tubes,

$$K = K_g r_{tot} + K_{ash} \omega_{ash} \qquad [5.13]$$

Here, $r_{tot} = r_{H_2O} + r_{RO_2}$ is the total volume fraction of triatomic gases in the flow of the external heat-transfer medium; r_{H_2O} is the volume fraction of water vapor; K_g is the radiant absorptance of triatomic gases:

$$K_g = \left(\frac{0.78 + 1.6 \cdot r_{H_2O}}{\sqrt{P \cdot r_n \cdot S}} - 1 \right)(1 - 0.37 \times 10^{-3} \cdot T) \qquad [5.14]$$

ω_{ash} is the concentration of ash particles in the gas flow; K_{ash} is the radiant absorptance of ash particles:

$$K_{ash} = \frac{10^4 \times B_{ash}}{\sqrt[3]{T^2_{s \cdot cont}}(1 + 1.2 \cdot \omega_{ash} \cdot S)} \qquad [5.15]$$

The coefficient B_{ash} is chosen from Table 5.1 depending on combustion products of which fuel make the gas flow, moving in the gas conduit.

Table 5.1 Values of empirical coefficient, B_{ash}

Fuel	B_{ash}
Anthracite	1.0
Black and lean coals	0.80
Brown coals	0.75
Shales	0.75
Peat	0.65

For a dust-free gas flow, the second term in Eq. [5.13] is taken to be zero.

$T_{s \cdot cont}$ is the segment-average temperature of the contaminated walls of tubes of the heating surface, determined from the value of Q_{seg}, assumed beforehand, using the equation:

$$T_{s \cdot cont} = \overline{T}_{seg} + \left[\beta_d \left(\frac{\delta_t}{\lambda_t} \cdot \frac{2}{1 + \beta_d} + \frac{1}{h_2} \right) + \varepsilon_{seg} \right] \frac{Q_{seg} \times 10^3}{A_{s \cdot tot}} + 273 \quad [5.16]$$

The obtained value of $T_{s \cdot cont}$ is refined if the values of Q_{seg}, assumed beforehand and found from Eq. [5.6], differ by more than 5%.

In Eq. [5.16], $\overline{T}_{seg}$ is the average temperature of the heated medium in the calculated segment, computed from the inlet T'_{seg} and outlet T''_{seg} temperatures of the medium:

$$\overline{T}_{seg} = \left(T'_{seg} + T''_{seg} \right) 0.5 \quad [5.17]$$

Since the inlet cross-sections of the calculated segment and considered element coincide, then the following:

$$T'_{seg} = T'$$

The outlet cross-section of the segment corresponds with the calculated cross-section of the considered element; therefore the temperature at the outlet from the calculated segment T''_{seg} is simultaneously the average temperature in the calculated cross-section of the element:

$$T''_{seg} = T_{c \cdot cr \cdot s}$$

T''_{seg} is determined according to Eqs. [5.4] and [5.5] using the value of Q_{seg}, assumed beforehand in Eq. [5.16] and, if needed, is refined together with $T_{s \cdot cont}$; $A_{s \cdot tot}$ is the total heating surface of the segment, calculated from the number of tubes z_{seg} within the segment according to Eqs. [2.2]−[2.5]; ε_{seg} is the contamination factor of tubes in the segment, $m^2 K/W$, determined according to Section 2.6 of Chapter 2; if the decrease in the heat absorption of the surface as a result of its contamination is taken into account by the thermal efficiency ξ, ε_{seg} can be approximately evaluated from the equation:

$$\varepsilon_{seg} = (1 - \xi)\left(\overline{\vartheta}_{seg} - \overline{T}_{seg} \right) \frac{A_{s \cdot tot}}{Q_{seg} \times 10^3} \quad [5.18]$$

where $\overline{\vartheta}_{seg}$ is the average gas temperature in the calculated segment, calculated from the inlet ϑ'_{seg} and the outlet ϑ''_{seg} gas temperatures in the segment:

$$\overline{\vartheta}_{seg} = 0.5\left(\vartheta'_{seg} + \vartheta''_{seg}\right) \qquad [5.19]$$

The gas temperature at the inlet to the calculated segment is taken to be the same as at the inlet to the considered element (the bundle):

$$\vartheta'_{seg} = \vartheta'$$

The gas temperature at the outlet from the calculated segment ϑ''_{seg} is approximately evaluated from the equation:

$$\vartheta''_{seg} = \vartheta' - (\vartheta' - \vartheta'')\frac{z_{2seg}}{z_2} \qquad [5.20]$$

in which ϑ'' is the gas temperature at the outlet from the considered bundle, z_2 is the number of tube rows in the bundle in the direction of the gas motion, z_{2seg} is the number of tube rows in the direction of the gas motion within the calculated segment, and $\overline{T}_{seg}$ is determined from Eq. [5.17].

5.2.2.2 Calculation of Heat Absorption by Convection
The heat absorption of the calculated segment by convection is determined from the expression:

$$Q_c = h_c \cdot A_{s \cdot tot}\left(\overline{\vartheta}_{seg} - t_{s \cdot cont}\right) \times 10^{-3}, \quad kW \qquad [5.21]$$

in which h_c is the convective heat transfer coefficient for the bundle on the whole, determined according to Section 2.44 of Chapter 2; $A_{s \cdot tot}$ is the total heating surface of the segment (Section 5.2.2.1); $\overline{\vartheta}_{seg}$ is the average gas temperature in the calculated segment, determined from Eq. [5.19]; and $t_{s \cdot cont}$ is the average temperature of the contaminated wall in the calculated segment:

$$t_{s \cdot cont} = T_{s \cdot cont} - 273 \qquad [5.22]$$

where $T_{s \cdot cont}$ is determined from Eq. [5.16].

5.2.3 Calculation of Medium Temperature Excess in the Maldistributed Tube

The excess of the medium temperature in the maldistributed tube over the average medium temperature in the calculated cross-section of the considered element is defined by the expression:

$$\Delta T_t = T_{max} - T_{c \cdot cr \cdot s} \qquad [5.23]$$

Here, $T_{c \cdot cr \cdot s}$ is the average temperature of the internal medium in the calculated cross-section of the element tubes, determined according to Section 5.2.1; and T_{max} is the medium temperature in the calculated cross-section of the most heated (maldistributed) tube of the element, determined from the enthalpy h_{max}, which is equal to this:

$$h_{max} = h_{c \cdot cr \cdot s} + \left(\frac{\eta_t \cdot \eta_{str}}{\rho_h} - 1 \right) \Delta h_x \qquad [5.24]$$

In Eq. [5.24], $h_{c \cdot cr \cdot s}$ is the medium enthalpy corresponding to the average temperature in the calculated cross-section of the element tubes $T_{c \cdot cr \cdot s}$, determined according to Section 5.2.1; Δh_x is the average increment of the medium enthalpy in the element up to the calculated cross-section, determined according to Section 5.2.2; η_{str} is the coefficient of structural nonequivalence of coils (the ratio of the heated surface of the maldistributed tube to the surface of the middle tube of the element, determined from geometric characteristics of tubes of the considered element); η_t is the coefficient of nonuniformity of heat absorption of the maldistributed tubes of the element based on the average heat absorption of the element.

For the gas temperature at the inlet to the bundle $\vartheta' \leq 900°C$, it is taken to be this:

$$\eta_t = 1.15,$$

irrespective of the position of the element over the width of the gas conduit; ρ_h is the coefficient of hydraulic maldistribution of the flow rate of the medium in the element tubes, which represents the ratio of the flow rates of the medium in the maldistributed tube to the mean flow rate of the medium in the element tubes; it is determined from the following equations:

- for steam superheaters in the general case

$$\rho_h = \sqrt{ \left(1 - \frac{\delta P_{col}}{\Delta P_{el}} \right) \frac{1}{\eta_h} \cdot \frac{\bar{v}}{\bar{v}_t} } \qquad [5.25]$$

- for steam superheaters with uniformly distributed supply and withdrawal of the medium with the same spatial arrangement of the inlet and outlet collectors

$$\rho_h = \sqrt{ \frac{1}{\eta_h} \cdot \frac{\bar{v}}{\bar{v}_t} } \qquad [5.26]$$

- for economizer surfaces, the values of ρ_h can be approximately taken to be $\rho_h \approx 0.8$ for steaming economizers, and $\rho_h \approx 0.9$ for nonsteaming economizers.

In Eqs. [5.25]–[5.26], δP_{col} is the difference of total pressure drops in the inlet and outlet collectors of the calculated element for cross-sections with the maldistributed tube and the mean flow rate of the medium, determined according to Section 4.6.5; ΔP_{el} is the total hydraulic resistance of the calculated element, determined according to the recommendations in Section 4.2; η_h is the coefficient of hydraulic nonuniformity, representing the ratio of the total coefficient of hydraulic resistance of the maldistributed tube z_t to the total coefficient of hydraulic resistance of the calculated element z_{el}:

$$\eta_h = \frac{z_t}{z_{el}} = \frac{\zeta_{fr}^t + \sum \zeta_{loc}^t}{\zeta_{fr}^{el} + \sum \zeta_{loc}^{el}} \qquad [5.27]$$

where ζ_{fr} and ζ_{loc} are determined according to the recommendations in Sections 4.4 and 4.5; v is the average specific volume of the medium in the element, determined from the average temperature and pressure of the medium in the calculated element; v_t is the average specific volume of the medium in the maldistributed tube, determined from the average temperature of the medium in the maldistributed tube $\overline{T}_t$:

$$\overline{T}_t = 0.5\left(T' + T_t''\right) \qquad [5.28]$$

Here, T' is the medium temperature at the inlet to the element; T_t'' is the medium temperature at the outlet from the maldistributed tube, which is determined with a subsequent refinement from the appropriate value of the enthalpy:

$$h_t'' = h' + \left(\frac{\eta_t \cdot \eta_{str}}{\rho_h} - 1\right)\overline{\Delta h}_{el} \qquad [5.29]$$

which is obtained from the preliminarily assumed value of ρ_h and the average increment of the medium enthalpy in the element $\overline{\Delta h}_{el}$; the value of $\frac{\overline{v}}{v_t}$ is taken to be unity for $\overline{\Delta h}_{el} \leq 160$ kJ/kg and medium pressures $P_f \leq 10$ MPa, and also for $\overline{\Delta h}_{el} \leq 120$ kJ/kg and $P_f > 10$ MPa.

If the inlet collector of the calculated element is intermediate and the admission of the medium into it does not provide its complete mixing, yet another term is introduced in the right-hand side of

Eq. [5.24] that takes account of the thermal maldistribution in the preceding element:

$$h_{max} = h_{c \cdot cr \cdot s} + \left(\frac{\eta_t \cdot \eta_{str}}{\rho_h} - 1 \right) \Delta h_x + a_{in \cdot m} \left(\frac{\eta_t - \eta_{str}}{\rho_h} - 1 \right)_{pr} \Delta h_{x_{pr}} \quad [5.30]$$

Here, the subscripts "pr" refers to the element preceding the calculated one; $a_{in \cdot m}$ is the coefficient taking account of the incompleteness of the medium mixing in the preceding element, which is taken to be the following:

- $a_{in \cdot m} = 0$ with a one-sided admission of the medium into the end of the collector or with a complete mixing of the medium in the preceding element;
- $a_{in \cdot m} = 0.5$ with a two-sided admission of the medium into the ends of the supply collector, and also with a small relative number of inlet tubes, distributed along the collector $n_{inl}/n_{out} \leq 0.3$;
- $a_{in \cdot m} = 1.0$ with an intermediate collector, and also with a large relative number of tubes, feeding the medium to the supply collector $n_{inl}/n_{out} > 0.3$.

5.2.4 Calculation of Specific Heat Load

The specific heat load at the point of maximum heat absorption of the most loaded tube is defined by the expression:

$$q_{max} = \eta_{wd} \cdot \eta_t \cdot q_0 \quad [5.31]$$

where η_{wd} is the coefficient of nonuniformity of heat absorption of the element across the width of the gas conduit, determined according to Section 5.2.2; η_t is the coefficient of nonuniformity of heat absorption of maldistributed tubes of the element across the width of the gas conduit, determined according to Section 5.2.3; q_0 is the average specific heat absorption of the most heated generating tube in the calculated cross-section, kW/m^2:

$$q_0 = \frac{\vartheta_{c \cdot r} - T_{c \cdot cr \cdot s}}{\beta_d \left(\frac{\delta_t}{k_t} \cdot \frac{2}{\beta_d + 1} + \frac{1}{h_2} \right) + \frac{1}{h_1} + 0.25 \cdot \varepsilon_{seg}} \times 10^{-3} \quad [5.32]$$

In Eq. [5.32], $\vartheta_{c \cdot r}$ is the gas temperature at the entrance to the calculated row, approximately evaluated from the equation:

$$\vartheta_{c \cdot r} = \vartheta' - (\vartheta' - \vartheta'') \frac{z_{2up \ c \cdot cr \cdot s}}{z_2} \quad [5.33]$$

where $z_{2\text{up c·cr·s}}$ is the number of tube rows in the direction of the gas motion up to the calculated cross-section; the other quantities are determined as in Eq. [5.20], Section 5.2.2.1; $T_{\text{c·cr·s}}$ is the average temperature of the heated medium in the calculated cross-section, determined according to Section 5.2.1; h_2 is the coefficient of heat transfer from the wall to the internal medium, determined according to Section 5.2.5; ε_{seg} is the contamination factor of tubes of the calculated segment, determined according to Section 5.2.2.1; h_1 is the coefficient of heat transfer from gases to the wall at the point of maximum heat absorption in the calculated cross-section, determined from the equation:

$$h_1 = \gamma_{\text{cfr}} \cdot h_{\text{1rdc}}^{\text{car}} + \varphi_{\text{p}} \cdot h_{\text{r}} \qquad [5.34]$$

where γ_{cfr} is the coefficient of nonuniformity of heat absorption over the circumference of the tube, taken to be $\gamma_{\text{cfr}} = 1.3$; $h_{\text{1rdc}}^{\text{car}}$ is the reduced coefficient of heat transfer from the gas side based on the surface of the finning-carrying tube A_{car}:

$$h_{\text{1rdc}}^{\text{car}} = h_{\text{c}} \left(\frac{A_{\text{car}}}{A} E \cdot \mu_{\text{r}} \cdot \psi_{\text{E}} + \frac{A_{\text{t}}}{A} \right) \qquad [5.35]$$

in this equation,

$$\frac{A_{\text{r}}}{A_{\text{car}}} = \psi_{\text{r}} - 1 - \frac{\delta_1}{s_{\text{r}}} \qquad [5.36]$$

$$\frac{A_{\text{t}}}{A_{\text{car}}} = 1 - \frac{\delta_1}{s_{\text{r}}} \qquad [5.37]$$

h_{c} is the circumference-average convective heat transfer coefficient in the calculated cross-section, determined according to Section 2.4.4 of Chapter 2 with account for the position of the calculated tube row over the bundle depth; here, the factor C_z in Eq. [2.28] should be determined as follows:

- for tubes of the first row, as for a three-row bundle
- for tubes of the second row, as for a five-row bundle
- and for tubes of the third and subsequent rows, $C_z = 1$

The other quantities in Eq. [5.35] are determined according to Section 2.3.1 of Chapter 2; h_{rad} is the heat transfer coefficient taking account of the heat, received through radiation from the gas volume preceding the bundle; it is allowed only for the first and the second rows of staggered bundles and for the first row of an in-line bundle and is obtained from the equation:

$$h_{\text{rad}} = \frac{q_{\text{rad}}}{\vartheta_{\text{c·r}} - T_{\text{s·cont}}} \times 10^3 \qquad [5.38]$$

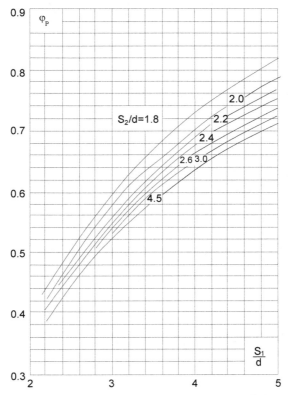

Figure 5.5 Irradiance coefficient of points with maximum specific heat absorption for tubes of staggered bundles.

Here, q_{rad} is determined from Eq. [5.9] in Section 5.2.2.1, $\vartheta_{c\cdot r}$ from Eq. [5.39] and $T_{s\cdot cont}$ from Eqs. [5.16]−[5.22].

φ_p is the irradiance coefficient of the point with maximum heat absorption, determined depending on the bundle arrangement and the number of the tube row with the calculated cross-section relative to the external radiation source as follows:

- For tubes of the first row behind the radiation source, $\varphi_p = 1$.
- For tubes of the second row of a staggered bundle, it is determined from Fig. 5.5.

The remaining designation in Eq. [5.32] is the same as in Eq. [5.1].

5.2.5 Determination of Convective Heat Transfer Coefficient

The convective heat transfer coefficient from the wall to the internal medium h_2 for non-boiling water and water vapor at pressures

$P_f \leq 18$ MPa, as well as for other uniphase heat-transfer media, is obtained from Eqs. [2.43] and [2.46] in Section 2.5.2 of Chapter 2. For boiling water, h_2 is determined from Fig. 5.6 as a function of the specific heat flux q_{max} based on the inside surface of the tube:

$$q_{max}^{in} = \beta_d \cdot \lambda \cdot q_{max} \qquad [5.39]$$

Here, h_2 is determined by the iteration method from q_{max}, evaluated beforehand using Eqs. [5.31] and [5.32].

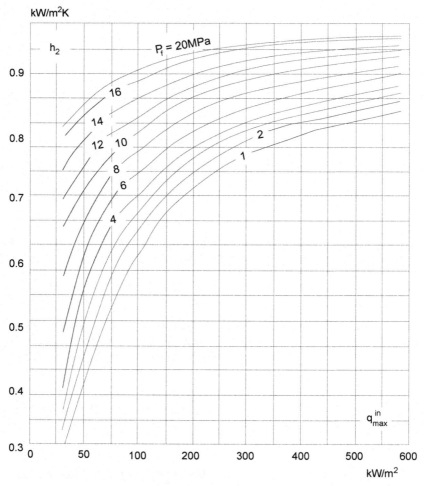

Figure 5.6 Heat transfer coefficient h_2 for boiling water.

5.3 TEMPERATURE AT FIN TIP

$$T_3 = \vartheta_{c \cdot r} - (\vartheta_{c \cdot r} - T_1) \frac{I_0(\beta_{r_e}) \cdot K_1(\beta_{r_e}) + I_1(\beta_{r_e}) \cdot K_0(\beta_{r_e})}{I_0(\beta_{r_1}) K_1(\beta_{r_e}) + I_1(\beta_{r_e}) \cdot K_0(\beta_{r_1})} \qquad [5.40]$$

where, T_1 is the temperature of the fin root, determined according to Section 5.2; I_0, I_1, K_0, and K_1 are modified Bessel functions, determined from the values of the arguments β_{r_1} and β_{r_e} according to Table 5.2. In the expressions for the arguments, β is the fin parameter (see Eq. [2.13] in Section 2.3.1 of Chapter 2); r_1 is the outside radius of the finning-carrying tube; r_e is the conventional finning radius, determined from the following equations:

- for circular fins,

$$r_e = \frac{D + \delta_r}{2} \qquad [5.41]$$

Table 5.2 Bessel functions

β_r	$I_0(\beta_r)$	$I_1(\beta_r)$	$K_0(\beta_r)$	$K_1(\beta_r)$
0.0	1.000	∞	0	∞
0.1	1.003	2.447	0.050	9.854
0.2	1.010	1.753	0.101	4.776
0.3	1.023	1.373	0.152	3.056
0.4	1.040	1.115	0.204	2.184
0.5	1.064	0.924	0.258	1.656
0.6	1.092	0.775	0.314	1.303
0.7	1.126	0.661	0.372	1.050
0.8	1.166	0.565	0.433	0.862
0.9	1.213	0.487	0.497	0.717
1.0	1.266	0.421	0.565	0.602
1.2	1.394	0.318	0.715	0.435
1.4	1.553	0.244	0.886	0.320
1.6	1.750	0.188	1.085	0.241
1.8	1.989	0.159	1.317	0.183
2.0	2.279	0.114	1.591	0.140
2.5	3.289	0.062	2.517	0.0739
3.0	4.881	0.0347	3.395	0.0402
3.5	7.378	0.0196	6.206	0.0222
4.0	11.302	0.0112	9.759	0.0125
4.5	17.481	0.0064	15.389	0.00708
5.0	27.240	0.0037	24.336	0.00404

- and for square fins,

$$r_e = \frac{c_{sq}\sqrt{2} + \delta_r}{2} \qquad [5.42]$$

5.4 MEAN INTEGRAL TEMPERATURE OF FIN

$$T_r = \vartheta_{c \cdot r} - (\vartheta_{c \cdot r} - T_1) \cdot E \qquad [5.43]$$

where, T_1 is determined according to Section 5.2, and E according to Section 2.3.1 of Chapter 2.

5.5 TEMPERATURE OF THE INSIDE SURFACE OF TUBE

$$T_2 = T_{c \cdot cr \cdot s} + \Delta T_t + \beta_d \cdot \lambda \cdot q_{max} \cdot \frac{1}{h_2} \qquad [5.44]$$

The quantities, entering into Eq. [5.44], are determined according to Section 5.2.

5.6 THICKNESS AVERAGE TEMPERATURE OF THE TUBE WALL IN THE REGION OF MAXIMUM HEAT FLUX

$$T_w = T_{c \cdot cr \cdot s} + \Delta T_t + \beta_d \cdot \lambda \cdot q_{max} \cdot \left(\Theta_k \cdot \frac{\delta_t}{k_t} \cdot \frac{1}{1 + \beta_d} + \frac{1}{h_2} \right) \qquad [5.45]$$

The quantities, entering into Eq. [5.45], are determined in accordance with Section 5.2.

CHAPTER 6

Strength Design

6.1 BASIC CONCEPTS

The strength design of heat-transfer surfaces composed of tubes with transverse finning lies, as a rule, in determining the allowable wall thickness (δ) of the finning-carrying tube, corresponding with the operating conditions and the selected material, and in comparing it with the nominal wall thickness δ_t, assumed beforehand in structural characteristics in the heat-transfer surface.

The design relies on such basic concepts as these:
- design pressure
- design temperature
- allowable stress

6.2 DESIGN PRESSURE

6.2.1 Definition of Design Pressure

The design pressure P is taken to mean the excess pressure of the working medium, on which the strength design of the given detail (the tube) is based.

The design pressure should be equal to or higher than the maximum pressure of the working medium that is possible for the given detail (the tube) under normal operating conditions. The need for the excess of the design pressure over the working pressure, as well as the magnitude of this excess, are determined with account for structural features of the heat-transfer device, parts it comes complete with (eg, safety valves), the purpose of the heat-transfer device, and experience in its operation.

6.2.2 Design Pressure Default Value

The design pressure in tubes of the heating surfaces is assumed to be equal to the pressure of the working medium at the entrance to the calculated bundle (the element).

In all cases, the design pressure should be taken to be at least 0.2 MPa.

Handbook for Transversely Finned Tube Heat Exchanger Design
ISBN 978-0-12-804397-4
http://dx.doi.org/10.1016/B978-0-12-804397-4.00006-9

6.3 DESIGN TEMPERATURE

6.3.1 Definition of Design Temperature

By the design wall temperature, T_w, is meant the metal temperature according to which the allowable stress is selected for the considered detail (the tube) of the heat-transfer device.

6.3.2 Calculation of Design Temperature

The temperatures on the outside T_1 and inside T_2 surfaces of the finned tube, the tube temperature average over the wall thickness T_w, and the temperature at the fin tip t are determined in accordance with the recommendations in Chapter 5.

According to the design temperature of the tube wall T_w, the heat-transfer surfaces are divided into two groups: low temperature and high temperature.

Among the low-temperature surfaces are surfaces with the maximum temperature of the tube walls lower than the temperature of the commencement of an intense creep of the metal. Among the high-temperature surfaces are surfaces for which the maximum temperature is higher than that indicated (Table 6.1).

The ultimate temperature of the outside tube surface corresponds with the temperature of the fin root T_1, determined from Eq. [5.1]. Its value should not be larger than those indicated in Table 6.2:

$$T_1 \leq [T_1] \tag{6.1}$$

Table 6.1 Classification of heat-transfer surfaces at low and high temperatures

Steel	Low-temperature surfaces	High-temperature surfaces
Carbon steel	$\leq 400°C$	$>400°C$
Alloy steel with reduced content of Cr and Mo	$\leq 425°C$	$>425°C$

Table 6.2 Allowable temperature of outside tube surface $[T_1]$

Steel grade	Carbon steel	Alloy steel with reduced content of Cr and Mo
$[T_1]$, °C	450	550

Table 6.3 Allowable temperature of fin metal [T_3]

Steel grade	Carbon steel	Alloy steel with reduced content of Cr and Mo
[T_3], °C	500	570

For the fin metal, the ultimate temperature T_3, calculated from Eq. [5.40], should not exceed the values indicated in Table 6.3:

$$T_3 \leq [T_3] \qquad [6.2]$$

6.4 ALLOWABLE STRESS

6.4.1 Definition of Allowable Stress

The nominal allowable stress [σ] is taken to mean the magnitude of stress used for determining the design thickness of the tube wall on the basis of the adopted initial data and the steel grade.

6.4.2 Determination of Allowable Stress Value

The nominal allowable stresses [σ] for rolled or wrought steel of grades, widely employed in convective finned heating surfaces, should be assumed in accordance with Tables 6.4—6.5 depending on the design temperature of the tube wall T_w and the service life of the heating surface.

For values of the service life intermediate to those indicated in the tables, it is permissible to determine the allowable stresses by the linear interpolation between the nearest values with a rounding off of up to 0.5 MPa to the side of decreasing magnitudes, if the difference between these values is no more than 20% of the average value of [σ] in the considered range. In other cases, the "logarithmic" interpolation should be used.

6.5 CALCULATION OF ALLOWABLE WALL THICKNESS OF FINNING-CARRYING TUBE

6.5.1 Basic Concepts

This calculation involves the following concepts:
- the design wall thickness δ_{calc}
- the nominal wall thickness δ_t
- the allowable wall thickness [δ]

Table 6.4 Nominal allowable stresses [σ, MPa] for carbon steels with reduced content of P and S

	Steel grade										
	0.08–0.12% C				0.15–0.16% C			0.18–0.2% C			
	Design service life, h										
T, °C	10^4	10^5	2×10^5	3×10^5	10^4	10^5	2×10^5	10^4	10^5	2×10^5	3×10^5
20–100	—	130	—	—	—	140	—	—	147	—	—
200	—	120	—	—	—	130	—	—	140	—	—
250	—	108	—	—	—	120	—	—	132	—	—
275	—	102	—	—	—	113	—	—	126	—	—
300	—	96	—	—	—	106	—	—	119	—	—
320	—	92	—	—	—	101	—	—	114	—	—
340	—	87	—	—	—	96	—	—	109	—	—
350	—	85	—	—	—	93	—	—	106	—	—
360	—	82	—	82	—	90	—	—	103	—	103
380	—	76	76	71	—	85	85	—	97	97	88
400	73	73	66	60	80	80	72	92	92	78	71
410	70	68	61	55	77	72	65	89	86	70	63
420	68	62	57	50	74	66	58	86	79	63	56
430	66	57	51	45	71	60	52	83	72	57	50
440	63	51	45	40	68	53	45	80	66	50	44
450	61	46	38	35	65	47	38	77	59	46	39
460	58	40	33	29	62	40	33	74	52	38	34
470	52	34	28	24	54	34	28	64	46	32	28
480	45	28	22	18	46	28	22	56	39	27	24
490	39	24	—	—	40	24	—	49	33	—	—
500	33	20	—	—	34	20	—	41	26	—	—
510	26	—	—	—	—	—	—	35	—	—	—

Table 6.5 Nominal allowable stresses [σ, MPa] for heat-resistant steels with content of Cr and Mo less than 1%

T, °C	Steel grade							
	0.12% C				0.15% C			
	Design service life, h							
	10^4	10^5	2×10^5	3×10^5	10^4	10^5	2×10^5	3×10^5
From 20 to 100	—	147	—	—	—	153	—	—
250	—	145	—	—	—	152	—	—
300	—	141	—	—	—	147	—	—
350	—	137	—	—	—	140	—	—
400	—	132	—	—	—	133	—	—
420	—	129	—	—	—	131	—	—
440	—	126	—	—	—	128	—	—
450	—	125	—	—	—	127	—	—
460	—	123	123	123	—	125	125	125
480	120	120	102	90	122	122	113	103
500	116	95	77	64	119	105	85	76
510	114	78	60	53	117	85	72	62
520	107	66	49	43	110	70	58	50

Continued

Table 6.5 Nominal allowable stresses [σ, MPa] for heat-resistant steels with content of Cr and Mo less than 1%—cont'd

| T, °C | 0.12% C | | | | 0.15% C | | | |
| | Design service life, h | | | | | | | |
	10^4	10^5	2×10^5	3×10^5	10^4	10^5	2×10^5	3×10^5
530	93	54	40	35	97	56	44	39
540	77	43			80	45	35	31
550	60				62	35	26	23
560					52	27		
570					42	21		
580								
590								
600								
610								
620								

Notes:
1. Underlined values are stresses determined from the yield stress as a function of temperature.
2. The allowable stresses in the columns for service lives of 10^4, 2×10^5 and 3×10^5 h, marked by the sign "—" above the bar, are taken to be equal to the values in the column for a service life of 10^5 h.
3. The allowable pressures with an overbar correspond to the operation of elements under the conditions of creep and are determined from the limit of the long-term strength for a pertinent service life.

6.5.2 Calculation of Design Wall Thickness

The design wall thickness δ_{calc} is calculated from the preset design pressure P and nominal allowable stress $[\sigma]$ using the equation:

$$\delta_{calc} = \frac{P \cdot d}{2\varphi_w[\sigma] + P} \qquad [6.3]$$

where d is the inside diameter of the tube, m; P is the design pressure, MPa, determined according to Section 6.2; $[\sigma]$ is the allowable stress, MPa, determined according to Section 6.4; and φ_w is the strength coefficient; for seamless tubes from which the finned heating surfaces are generally made:

$$\varphi_w = 1 \qquad [6.4]$$

6.5.3 Calculation of Allowable Wall Thickness

The allowable wall thickness $[\delta]$ is determined from the design wall thickness with account for the addition term C:

$$[\delta] = \delta_{calc} + C \qquad [6.5]$$

The value of C is assumed in accordance with Section 6.5.4.

6.5.4 Calculation of the Additive to the Design Wall Thickness

The additive to the design wall thickness C is made up of the manufacturing C_1 and operational C_2 additives

$$C = C_1 + C_2 \qquad [6.6]$$

C_1 is determined according to Section 6.5.5, and C_2 according to Section 6.5.6.

6.5.5 Calculation of the Manufacturing Additive

The manufacturing additive C_1 consists of additives that make up for a possible decrease in the strength of the detail during its fabrication due to a minus deviation of the wall thickness of the semifinished product C_{11} and to technological thinnings C_{12}:

$$C_1 = C_{11} + C_{12} \qquad [6.7]$$

The value of the additive C_{11} should be determined from the limiting minus deviation of the wall thickness of tubes of the preselected dimension type, established by standards or specifications. If such data are not available, the following can be assumed:

$$C_{11} = 0.15 \cdot \delta_t \qquad [6.8]$$

The value of the additive C_{12} is determined by the technology of manufacturing the detail and assumed according to relevant technical conditions. For seamless steel tubes, the following can be assumed:

$$C_{12} = 0 \qquad [6.9]$$

6.5.6 Calculation of the Operational Additive

The operational additive consists of additives that make up for a decrease in the tube strength under operational conditions due to corrosion and erosion from the side of the internal medium C_{21} and from the side of the external medium C_{22}:

$$C_2 = C_{21} + C_{22} \qquad [6.10]$$

The value of the additive C_{21} for tubes made of carbon and low–alloy heat–resistant steels with a design service life of 10^5 h is determined from Table 6.6 according to the outside diameter of tubes d and the type of the internal medium.

The value of the additive C_{22} in the general case depends on the temperature of the outside surface of the tube T_1, the composition of the external heat-transfer medium (combustion products), and the type of metal (the grade of steel) of which the tube is made. To determine the additive C_{22}, the temperature of the outside surface of the tube t_1 should be compared with the allowable temperature $[T_1]$, whose values for carbon and low-alloy heat-resistant steels are presented in Table 6.2 with reference to combustion products of the main power-generating fuels.

Table 6.6 Values of the additive C_{21}

Working medium	$D \leq 32$ mm	32 mm $< d \leq 76$ mm
Water, steam–water mixture, dry saturated steam	0	0.5 mm
Superheated steam	0	0.3 mm

The value of the additive C_{22} for the service life of the heating service equal to 10^5 h should be assumed minimum, proceeding from the conditions:

$$\text{for } T_1 < \{[T_1] - 40°C\} \quad C_1 + C_2 \geq 0.5 \text{ mm;} \qquad [6.11]$$

$$\text{for } \{[T_1] - 40°C\} < T_1 \leq [T_1] \quad C_1 + C_2 \geq 1.0 \text{ mm.} \qquad [6.12]$$

For example, if $T_1 < \{[T_1] - 40°C\}$, $C_{11} = 0.3$ mm, $C_{12} = 0$ mm, and $C_{21} = 0.5$ mm, then $C_{11} + C_{12} + C_{21} = 0.8$ mm, and for condition [6.11] to be fulfilled, the additive C_{22} should be taken to be zero ($C_{22} = 0$). If at the same values of T_1 and $[T_1]$ $C_{11} = 0.3$ mm, $C_{12} = 0$ mm, and $C_{21} = 0$ mm, then $C_{11} + C_{12} + C_{21} = 0.3$ mm, and for condition [6.11] to be fulfilled, it is necessary to assume $C_{22} = 0.2$ mm.

The value of the additive C_{22} for general-purpose tubes made of carbon steel should be at least 0.4 mm, regardless of the surface temperature, steel grade, and quality.

For a service life shorter than 10^5 h, it is admissible to decrease the value of the additive C_{22} in proportion to the ratio of a given service life to that equal to 10^5 h.

6.5.7 Calculated Finned Heating Surface

The value of the allowable wall thickness $[\delta]$, obtained in accordance with Sections 6.5.2–6.5.6, is compared with the preliminarily adopted nominal value of the wall thickness of tubes of the calculated finned heating surface δ_t. The nominal wall thickness should not be smaller than the allowable value,

$$\delta_t \geq [\delta]. \qquad [6.13]$$

CHAPTER 7

Examples of Calculations

7.1 CALCULATION ASSIGNMENT

To perform the designing of a Water-to-Air Heat Exchanger (WAHE) for a Turbo-Refrigerating Unit (TRU) with the following characteristics:

Air temperature at the inlet	$\vartheta' = 95°C$
Air temperature at the outlet	$\vartheta'' = 40°C$
Air pressure at the inlet	$P'_{aire} = 0.2 \text{ MPa}$
Total air pressure loss in WAHE	$\Delta h \leq 0.001 \text{ MPa}$
Temperature of cooling water at WAHE inlet	$T' = 25°C$
Air flow rate	$G_g = 2.7 \text{ kg/s}$
Water flow rate	$G_f = 3.0 \text{ kg/s}$
Water pressure at inlet	$P'_f = 0.5 \text{ MPa}$
Hydraulic resistance of WAHE water conduit	$\Delta P_{el} \leq 0.01 \text{ MPa}$
Overall dimensions of gas conduit $(a \times b \times c)$, no larger than	$0.56 \times 0.5 \times 1.0 \text{ m}$

7.2 HEAT BALANCE

The thermal power of WAHE is determined from the equation:

$$Q = G_g c_p (\vartheta' - \vartheta'')$$

where c_p is the specific heat of air at $P_g = 0.2$ MPa in the temperature range $\vartheta = 40-95°C$.

According to Standards [12], in the aforementioned range,

$$C_P \approx \text{const} = 1.010 (\text{kJ/kg K})$$

as a result,

$$Q = 2.7 \cdot 1.010(95 - 40) = 150.3 \text{ kW}.$$

From the equation of heat balance,

$$Q = G_f \cdot (h'' - h'),$$

Handbook for Transversely Finned Tube Heat Exchanger Design
ISBN 978-0-12-804397-4
http://dx.doi.org/10.1016/B978-0-12-804397-4.00007-0

the water enthalpy and temperature at the WAHE outlet are determined as follows. In accordance with Table S.2, on the basis of the inlet pressure $P'_m = 0.5$ MPa and the inlet temperature $T' = 25°C$, the water enthalpy at the WAHE inlet is assumed:

$$h' = 105.1 \ (\text{kJ/kg})$$

then,

$$h'' = h' + \frac{Q}{G_f} = 105.1 + \frac{150.3}{3.0} = 155.2 \ (\text{kJ/kg})$$

in accordance with Table S.2,

$$T'' = 37.0°C.$$

The average water temperature in WAHE is shown:

$$\overline{T} = 0.5(T' + T'') = 0.5(25 + 37) = 31°C.$$

The average air temperature in WAHE is as follows:

$$\overline{\vartheta} = 0.5(\vartheta' + \vartheta'') = 0.5(95 + 40) = 67.5°C.$$

The physical properties of water at $P'_m = 0.5$ MPa and $T = 31°C$ (according to Lokshin et al. [9]) are as follows:

Density	$\rho_f = 1000$ kg/m^3
Specific volume	$v_f = 0.001$ m^3/kg
Kinematic viscosity	$\nu_f = 0.783 \times 10^{-6}$ m^2/s
Thermal conductivity	$k_f = 0.619$ W/m K
Prandtl number	$\text{Pr}_f = 5.292$

The physical properties of air at $P_m = 0.2$ MPa and $\vartheta = 67.5°C$ (according to Standards [12]) are as follows:

Density	$\rho_g = 2.049$ kg/m^3
Specific volume	$v_g = 0.488$ m^3/kg
Kinematic viscosity	$\nu_g = 0.9945 \times 10^{-5}$ m^2/s
Specific heat	$C_p = 1.010$ kJ/kg K
Thermal conductivity	$k_g = 0.0294$ W/m K
Thermal diffusivity	$\alpha_g = 1.416 \times 10^{-5}$ m^2/s
Prandtl number	$\text{Pr}_g = 0.702$

7.3 STRUCTURAL FEATURES OF WAHE

WAHE is designed as a staggered bundle of tubes with external helical finning. The tubes are joined into coils longitudinal with respect to the air passing, whose bends are located outside the gas conduit (Fig. 7.1).

Bimetallic tubes with the following characteristics are selected:
The internal tube:

Material	Carbon steel used in steam generators
Inside diameter	$d_{in} = 0.020$ m
Wall thickness	$\delta'_t = 2.5 \times 10^{-3}$ m

The external tube with rolled helical finning:

Material	Aluminum
Diameter of the finning-carrying tube	$d_{in} = 0.028$ m
Fin height	$l_r = 0.0135$ m
Fin spacing	$s_r = 0.003$ m
Average fin thickness	$\delta_r = 8 \times 10^{-4}$ m
Thermal contact resistance on steel–aluminum interface	$R_{cont} = 1.89 \times 10^{-4}$ m^2 K/W

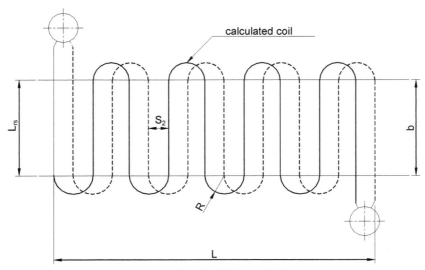

Figure 7.1 Sketch of designed WAHE.

7.3.1 Specific Geometric Characteristics of Finned Tubes

Specific characteristics of finned tubes in the considered case are determined taking into account that the coil bends will be located outside the gas conduit, and as a consequence,

$$A_{\text{total}} = A.$$

The surface area of fins per 1 m of the tube length (according to Eq. [2.3]) is shown:

$$
\begin{aligned}
A_{r1} &= \frac{\pi}{2}(D^2 - d^2 + 2D\delta_r)\frac{L_{rs}}{S_r}z \\
&= \frac{\pi}{2}(0.055^2 - 0.028^2 + 2 \cdot 0.055 \cdot 0.0008)\frac{1}{0.003}1 = 1.219 \text{ m}^2.
\end{aligned}
$$

The surface area of the carrying tube, not occupied by fins, per 1 m of the length of the finned tube is shown next:

$$
\begin{aligned}
A_{t1} &= \pi \cdot d\left[L_{rs}\left(1 - \frac{\delta_r}{S_r}\right)z + L_{\tau}\right] \\
&= \pi \cdot 0.028\left[1\left(1 - \frac{0.0008}{0.003}\right)1 + 0\right] = 0.0645 \text{ m}^2.
\end{aligned}
$$

The area of the outside surface per 1 m of the tube length is calculated:

$$A_1 = A_{r1} + A_{t1} = 1.219 + 0.0645 = 1.284 \text{ m}^2.$$

The ratio A_r/A is as follows:

$$\frac{A_r}{A} = \frac{A_{r1}}{A_1} = \frac{1.219}{1.284} = 0.95.$$

The ratio A_t/A is as follows:

$$\frac{A_t}{A} = \frac{A_{t1}}{A_1} = \frac{0.0645}{1.284} = 0.05.$$

The total surface area of the finning-carrying tube per 1 m of the tube length is calculated:

$$A_{rct1} = \pi \cdot d \cdot L_{\tau} = \pi \cdot 0.028 \cdot 1 = 0.088 \text{ m}^2.$$

The area of the inside surface per 1 m of the tube length is shown:

$$A_{in1} = \pi \cdot d_{in} \cdot L_{\tau} = \pi \cdot 0.020 \cdot 1 = 0.0628 \text{ m}^2.$$

The ratio A/A_{in} is as follows:

$$\frac{A}{A_{in}} = \frac{A_1}{A_{in1}} = \frac{1.284}{0.0628} = 20.45.$$

The fin coefficient Ψ_r follows:

$$\Psi_r = \frac{A}{A_{rct}} = \frac{A_1}{A_{rct1}} = \frac{1.284}{0.088} = 14.6.$$

7.3.2 Dimensions of Gas Conduit and Spacing Characteristics of WAHE

Taking into account the limitations on overall dimensions of the heat exchanger, we select a maximally dense "equilateral" arrangement of the bundle, with which the longitudinal and the transverse tube spacings are related as shown:

$$S_2 = \frac{\sqrt{3}}{2} S_1. \tag{7.1}$$

We also assume dimensions of the gas conduit guided by the specified limitations on size:
- width of gas circuit $a = 0.56$ m
- height of gas conduit $b = 0.5$ m

The transverse tube spacing S_1 is chosen with account for the maximum transverse dimension of the finned tube, ie, the outside diameter of finning:

$$D = d + 2 \cdot l_r = 0.028 + 2 \cdot 0.0135 = 0.055 \text{ m}$$

and also for the need for the observance of equality of the number of tubes z_1 in all transverse rows of the bundle:

$$S_1 = \frac{a}{z_1 + 0.5} = \frac{0.560}{9.5} = 0.059 \text{ m} > D = 0.055 \text{ m};$$

here, $z_1 = 9$ is the maximum number of finned tubes that fit into the width $n = 0.56$ m with a sufficient technological clearance.

With allowance for relation Eq. [7.1], the longitudinal tube spacing is calculated:

$$S_2 = \frac{\sqrt{3}}{2} 0.059 = 0.051 \text{ m}.$$

The diagonal tube spacing with an "equivalent" arrangement is equal to the transverse tube spacing:

$$S_2' = S_1 = 0.059 \text{ m}$$

Thus, we finally assume the following:
the transverse tube spacing $S_1 = 0.059$ m
the longitudinal tube spacing $S_2 = 0.051$ m
the diagonal tube spacing $S_2' = 0.059$ m.
The relative spacing characteristics are shown:

$$\sigma_1 = \frac{S_1}{d} = \frac{0.059}{0.028} = 2.11; \quad \sigma_2 = \frac{S_2}{d} = \frac{0.051}{0.028} = 1.82;$$

$$\sigma_2' = \frac{S_2'}{d} = \frac{0.059}{0.028} = 2.11; \quad \frac{\sigma_1}{\sigma_2} = \frac{S_1}{S_2} = 1.16.$$

7.3.3 Free Area for Passage of Air and Calculated Air Velocity

In accordance with Section 2.4.2 of Chapter 2, to determine the minimum free area for the passage of the external heat transfer medium with a staggered arrangement of tubes, we find the conventional diameter of the finned tube (Eq. [2.22]):

$$d_{cl} = d + \frac{2 \cdot l_r \cdot \delta_r}{s_r} = 0.028 + \frac{2 \cdot 0.0135 \cdot 8 \times 10^{-4}}{3 \times 10^{-3}} = 0.0352 \text{ m}$$

and afterward the bundle diameter (Eq. [2.21]):

$$\varphi_{cl} = \frac{S_1 - d_{cl}}{S_2' - d_{cl}} = \frac{0.059 - 0.0352}{0.059 - 0.0352} = 1.$$

For $\varphi_{cl} \leq 2$, the minimum free area is situated in the plane of transverse spacing and is equal to (Eq. [2.23]):

$$F = a \cdot b - z_1 \cdot L_{c \cdot cr \cdot s} \cdot d_{cl} = 0.56 \cdot 0.5 - 9 \cdot 0.5 \cdot 0.0352 = 0.1216 \text{ m}^2$$

Here, the length of tubes within the confines of the gas conduit is taken to be equal to its height:

$$L_{c \cdot cr \cdot s} = b = 0.5 \text{ m}$$

According to Section 7.2, the average specific volume of air at $P_g = 0.2$ MPa and $\vartheta = 67.5°C$ is this:

$$v_g = 0.488 \text{ m}^3/\text{kg}$$

The calculated air velocity is shown next:

$$u_g = \frac{G_g \cdot v_g}{F} = \frac{2.7 \cdot 0.488}{0.1216} = 10.84 \text{ m/s}$$

7.3.4 Free Area for Passage of Water and Average Water Velocity

In accordance with Section 2.5.2 of Chapter 2, the free area for the passage of the internal heat-transfer medium is obtained from Eq. [2.48], in which the number of tubes, connected in parallel, z_{tcp} is determined by the number of tubes in the transverse row of the bundle z_1 and by the number of the coil starts n_x. With a staggered arrangement of tubes,

$$n_x \geq 2;$$

we assume $n_x = 2$, then,

$$z_{tcp} = n_x \cdot z_1 = 2 \cdot 9 = 18$$

$$f = z_{tcp} \cdot \frac{\pi \cdot d_{in}^2}{4} = 18 \cdot \frac{\pi \cdot 0.02^2}{4} = 0.00566 \text{ m}^2.$$

According to Eq. [2.47], the average water velocity is this:

$$u_f = \frac{G_f \cdot v_f}{f} = \frac{3.0 \cdot 0.001}{0.00566} = 0.53 \text{ m/s}.$$

7.4 AREA OF HEAT-TRANSFER SURFACE

7.4.1 Calculation of Area of Heat-Transfer Surface of WAHE

In accordance with Section 2.1.1 of Chapter 2, the area of the heat-transfer surface of WAHE can be found from Eq. [2.1]:

$$A = \frac{Q \times 10^3}{U \cdot \Delta T}.$$

For which purpose it is necessary to calculate the overall heat transfer coefficient U and the average temperature difference ΔT.

7.4.2 Overall Heat Transfer Coefficient

According to Section 2.2.1 of Chapter 2, the overall heat transfer coefficient should be determined from Eq. [2.6]:

$$U = \frac{\Psi}{\frac{A}{A_{in}} \cdot \frac{1}{h_2} + \frac{A}{A_{in}} \cdot R_T + \frac{1}{h_{1rdc}}}.$$

In this equation, according to Section 2.6.6 of Chapter 2, the thermal efficiency Ψ (the heat-transfer medium is clean air) is taken to be the following:

$$\Psi = 0.95.$$

In determining the thermal resistance of the wall of bimetallic tubes, it is possible to disregard the quantities $\frac{\delta'_T}{k'_T}$ and $\frac{\delta''_T}{k''_T}$ owing to their smallness and to assume $R_T \approx R_{cont}$. According to the data sheet, the thermal contact resistance is $R_{cont} \approx 1.89 \times 10^{-4} \, \mathrm{m^2 \, K/W}$. Thus, the following can be assumed:

$$R_T = R_{cont} = 1.89 \times 10^{-4} \, \mathrm{m^2 \, K/W}.$$

Following Section 7.3.1 of the present calculation, the ratio $\frac{A}{A_{in}}$ is shown:

$$\frac{A}{A_{in}} = 20.45.$$

7.4.3 Reduced Heat Transfer Coefficient h_{1rdc}

According to Section 2.3.1 of Chapter 2, when finned tubes are washed by the flow of a clean heat-transfer medium (air), h_{1rdc} should be determined from Eq. [2.10]:

$$h_{1rdc} = \left(\frac{A_r}{A} \cdot E \cdot \mu_r \cdot \psi_E + \frac{A_t}{A} \right) h_c$$

The calculation of h_{1rdc} should be started with determining the convective heat transfer coefficient h_c, since E and ψ_E are also functions of h_c.

1. **The convective heat transfer coefficient h_c**

   In accordance with Section 2.4.4 of Chapter 2, the shape parameter of the bundle X is determined, which for a staggered arrangement of tubes is (Eq. [2.31]):

   $$X = \frac{\sigma_1}{\sigma_2} \cdot \frac{1.26}{\Psi_r} - 2 = 1.16 - \frac{1.26}{14.6} - 2 = -0.926.$$

   The exponent n and the factor C_q in Eq. [2.28] are obtained from Eqs. [2.29] and [2.30], respectively:

   $$n = 0.7 + 0.08 \cdot \tanh(X) + 0.005 \cdot \Psi_r$$

   $$= 0.7 + 0.08 \cdot \tanh(-0.926) + 0.005 \cdot 14.6 = 0.715;$$

   $$C_q = (1.36 - \tanh(X)) \cdot \left(\frac{1.1}{\Psi_r + 8} - 0.014 \right)$$

   $$= (1.36 - \tanh(-0.926)) \cdot \left(\frac{1.1}{14.6 + 8} - 0.014 \right) = 0.0725.$$

   The number of transverse tube rows in the bundle is taken to be $z_2 > 10$. Then, the factor C_z in Eq. [2.28] is this:

   $$C_z = 1.0.$$

   Substituting the obtained values of m, C_q, and C_z, as well as the physical properties of air from Section 7.2 of the present calculation, into Eq. [2.28] yields the following value of h_c:

   $$h_c = 1.13 \cdot C_z \cdot C_q \cdot \frac{k_g}{d} \cdot \left(\frac{u_g \cdot d}{v_g} \right)^n \cdot Pr_g^{0.33}$$

   $$= 1.13 \cdot 1.0 \cdot 0.0725 \cdot \frac{0.0294}{0.028} \cdot \left(\frac{10.84 \cdot 0.028}{0.9945 \times 10^{-5}} \right)^{0.715} \cdot (0.702)^{0.33}$$

   $$= 123.1 \ W/m^2 \ K$$

2. **The true efficiency of the fin**

   According to Section 2.3.1 of Chapter 2, the true efficiency of the fin is defined by the product $E \cdot \mu_r \cdot \Psi_E$.

   For determining the first multiplier, ie, the theoretical efficiency of the fin, the fin parameter β needs to be calculated (Eq. [2.13]). For this, in turn, it is necessary to know the thermal conductivity of the fin at its average temperature T_r. Let us evaluate T_r according to Eq. [2.16], assuming with a possible subsequent refinement $E' \approx 0.9$:

$$T_r = \vartheta - (\vartheta - T)\cdot E' = 67.5 - (67.5 - 31)\cdot 0.9 \approx 28°C;$$

at $T_r = 28°C$, the thermal conductivity of aluminum is $k_r = 225$ W/m K. Then,

$$\beta = \sqrt{\frac{2h_c}{\delta_r \cdot k_r}} = \sqrt{\frac{2\cdot 123.1}{0.0008\cdot 225}} = 36.98 \text{ m}^{-1}.$$

The conventional fin height l'_r is determined from Eq. [2.14]:

$$l'_r = l_r \cdot \left[1 + \left(0.191 + 0.054\frac{D}{d}\right)\cdot \ln\left(\frac{D}{d}\right)\right]$$

$$= 0.0135 \cdot \left[1 + \left(0.191 + 0.054\frac{0.055}{0.028}\right)\cdot \ln\left(\frac{0.055}{0.028}\right)\right] = 0.0162 \text{ m}.$$

The theoretical efficiency of the fin is calculated:

$$E = \frac{\tanh\left(\beta l'_r\right)}{\beta l'_r} = \frac{\tanh(36.98\cdot 0.0162)}{36.98\cdot 0.0162} = 0.895.$$

Evidently, there is no need for further refinement of the value of k_r. The correction factor ψ_E is determined from Eq. [2.15]:

$$\psi_E = 1 - 0.016\cdot \left(\frac{D}{d} - 1\right)[1 + \tanh(2\beta l_r - 1)]$$

$$= 1 - 0.016\cdot \left(\frac{0.055}{0.028} - 1\right)[1 + \tanh(2\cdot 36.98\cdot 0.0135 - 1)] = 0.985.$$

The coefficient μ_r for fins with an approximately constant thickness is taken to be $\mu_r = 1.0$ (Fig. 2.1).

Substituting the obtained values of h_c, E, μ_r and ψ_E, as well as the values of A_r/H and A_t/H from Section 7.3.1 of the present calculation into Eq. [2.10] gives the reduced heat transfer coefficient:

$$h_{1rdc} = \left(\frac{A_r}{A}\cdot E\cdot \mu_r\cdot \psi_E + \frac{A_r}{A}\right)\cdot h_c = (0.95\cdot 0.895\cdot 1.0\cdot 0.985 + 0.05)\cdot 123.1$$

$$= 109.2 \text{ W/m}^2 \text{ K}$$

7.4.4 Coefficient of Heat Transfer From Wall to Internal Medium h_2

According to Section 2.5.2 of Chapter 2, to determine h_2, it is necessary to know the Reynolds (Re_m) and Prandtl (Pr_m) numbers. According to Sections 7.2 and 7.3.4, $Pr_f = 5.292$, $\nu_f = 0.783 \times 10^{-6}\ \text{m}^2/\text{s}$ and $u_f = 0.53\ \text{m/s}$. Then,

$$Re_f = \frac{u_f \cdot d_{in}}{\nu_f} = \frac{0.53 \cdot 0.02}{0.7830 \times 10^{-6}} = 13,538.$$

At such Re_f and Pr_f numbers, the heat transfer coefficient h_2 should be calculated from Eq. [2.43]. For this purpose, the quantities λ (Eq. [2.44]) and ζ (Eq. [2.45]) are determined that enter into expression [2.43]:

$$\lambda = 1 + \frac{900}{Re_f} = 1 + \frac{900}{13,538} = 1.066$$

$$\zeta = (1.82 \cdot \lg(Re_f) - 1.64)^{-2} = (1.82 \cdot \lg(13,538) - 1.64)^{-2} = 0.0289.$$

To determine the correction C_{tem} in Eq. [2.43], the average temperature of the inside surface of the tube T_{in} is evaluated assuming, with a subsequent refinement, the area of the outside heat transfer surface $A' = 100\ \text{m}^2$ and the heat transfer coefficient $h_2' = 3500\ \text{W/m}^2\ \text{K}$. Then, with account for Section 7.3.2 and Eq. [2.50a],

$$A'_{in} = \frac{A'}{(A/A_{in})} = \frac{100}{20.45} = 4.9\ \text{m}^2$$

$$T'_{in} = T + \frac{Q}{A'_{in}} \cdot \frac{10^3}{h_2'} = 31 + \frac{150.3}{4.9} \cdot \frac{10^3}{3500} = 31 + 9.0 = 40°C$$

for the considered case where the internal heat-transfer medium is dropping liquid (water), and it is heated (the heat flux is directed from the tube wall to water), the correction C_{tem} is determined from the equation:

$$C_{tem} = \left(\frac{\mu_f}{\mu_{in}} \right)^{0.11}.$$

The dynamic viscosity of water at $P'_f = 0.5\ \text{MPa}$ and $T_f = T = 31°C$ (according to Lokshin et al. [9]) is shown:

$$\mu_f = 0.783 \times 10^{-3}\ \text{Pa s};$$

the dynamic viscosity of water at $P_f' = 0.5$ MPa and $T_{in} = T' = 40°C$ [9] is

$$\mu_{in} = 0.6510 \times 10^{-3} \text{ Pa s};$$

in view of this,

$$C_{tem} = \left(\frac{0.7830 \times 10^{-3}}{0.6510 \times 10^{-3}}\right)^{0.11} = 1.021.$$

Next, the values of the obtained quantities are substituted into Eq. [2.43]:

$$
\begin{aligned}
h_2 &= \frac{k_f}{d_{in}} \cdot \left[\frac{0.125 \cdot \zeta \cdot Re_f \cdot Pr_f \cdot C_{tem}}{\lambda + 4.5 \cdot \zeta^{0.5}\left(Pr_f^{0.666} - 1\right)}\right] \\
&= \frac{0.619}{0.020} \cdot \left[\frac{0.125 \cdot 0.0289 \cdot 13,538 \cdot 5.292 \cdot 1.021}{1.066 + 4.5 \cdot 0.0289^{0.5}\left(5.292^{0.666} - 1\right)}\right] = 3119.7 \text{ W/m}^2 \text{ K}.
\end{aligned}
$$

Returning to Section 7.4.2, we determine the overall heat transfer coefficient:

$$
\begin{aligned}
U &= \frac{\Psi}{\frac{A}{A_{in}} \cdot \frac{1}{h_2} + \frac{A}{A_{in}} \cdot R_T + \frac{1}{h_{1rdc}}} = \frac{0.95}{20.45 \cdot \frac{1}{3119.7} + 20.45 \cdot 1.89 \times 10^{-4} + \frac{1}{109.2}} \\
&= 48.51 \text{ W/m}^2 \text{ K}.
\end{aligned}
$$

7.4.5 Average Temperature Difference Δt

We assume a countercurrent scheme of motion of the heat-exchanging media. According to Section 2.7.2 of Chapter 2, in this case Δt is calculated from Eq. [2.53]:

$$\Delta T_1 = \vartheta' - T'' = 95 - 37 = 58.0°C;$$
$$\Delta T_{sm} = \vartheta'' - T' = 40 - 25 = 15.0°C.$$

As a result,

$$\Delta T = \frac{\Delta T_1 - \Delta T_{sm}}{\ln\left(\frac{\Delta T_1}{\Delta T_{sm}}\right)} = \frac{58.0 - 15.0}{\ln\left(\frac{58.0}{15.0}\right)} = 31.8(°C).$$

7.4.6 Results of Thermal Calculations

1. Returning to Section 7.4.1, we determine the area of the heat transfer surface of WAHE:

$$A = \frac{Q \times 10^3}{U \cdot \Delta T} = \frac{150.3 \times 10^3}{48.51 \cdot 31.8} = 97.43 \text{ m}^2.$$

2. The value of the correction C_{tem} in the equation for h_2 (Section 7.4.4) is refined as follows:

$$A_{in} = \frac{A}{\left(\dfrac{A}{A_{in}}\right)} = \frac{97.43}{20.45} = 4.76 \text{ m}^2$$

$$T_{in} = T + \frac{Q \times 10^3}{A_{in} \cdot h_2} = 31 + \frac{150.3}{4.76} \cdot \frac{10^3}{3119.7} = 41.0°\text{C}.$$

As seen, $T_{in} \approx T'_{in}$, and there is no need for refining the value of C_{tem}: $C_{tem} = 1.021$.

3. The total length of finned tubes (with no bends) of WAHE is determined:

$$L_{rs} = \frac{A}{A_1} = \frac{97.43}{1.284} = 75.9 \text{ m}.$$

4. The total number of tubes in WAHE is calculated:

$$z = \frac{L_{rs}}{L_{c \cdot cr \cdot s}} = \frac{75.9}{0.5} = 152.$$

5. The number of transverse rows of tubes is determined:

$$z_2 = \frac{z}{z_1} = \frac{152}{9} = 17.$$

We assume $z_2 = 18$.

6. The depth of the gas conduit is shown:

$$c = (z_2 - 1) \cdot S_2 = (18 - 1) \cdot 0.051 = 0.87 \text{ m}.$$

7. The actual number of tubes in WAHE is calculated:

$$z^a = z_1 \cdot z_2 = 9 \cdot 18 = 162.$$

8. The actual length of finned tubes of WAHE is determined:

$$L_{rs}^a = L_{c \cdot cr \cdot s} \cdot z^a = 0.5 \cdot 162 = 81 \text{ m}.$$

7.5 CALCULATION OF AERODYNAMIC RESISTANCE

7.5.1 Determination of Length of Developed Surface and Diameter of Bundle

According to Section 3.1.1 of Chapter 3, the calculation of the aerodynamic resistance should start with determining the reduced length of the developed surface $\frac{A_{\text{total}}}{F}$ and the equivalent diameter of the most contracted cross-section of the bundle d_{eq}.

The quantity $\frac{A_{\text{total}}}{F}$ is calculated from Eq. [3.9]:

$$
\begin{aligned}
\frac{A_{\text{total}}}{F} &= \frac{\pi \cdot [d \cdot s_{\text{r}} + 2 \cdot l_{\text{r}} \cdot \delta_{\text{r}} + 2 \cdot l_{\text{r}} \cdot (l_{\text{r}} + d)]}{S_1 \cdot s_{\text{r}} - (d \cdot s_{\text{r}} + 2 \cdot l_{\text{r}} \cdot \delta_{\text{r}})} \\
&= \frac{\pi \cdot [0.028 \cdot 0.003 + 2 \cdot 0.0135 \cdot 0.0008 + 2 \cdot 0.0135 \cdot (0.0135 + 0.028)]}{0.059 \cdot 0.003 - (0.028 \cdot 0.003 + 2 \cdot 0.0135 \cdot 0.0008)} \\
&= 53.95.
\end{aligned}
$$

The equivalent diameter for staggered bundles of tubes with $\varphi_{\text{cl}} \leq 2$ is determined from Eq. [3.10]:

$$
\begin{aligned}
d_{\text{eq}} &= \frac{2[s_{\text{r}}(S_1 - d) - 2l_{\text{r}}\delta_{\text{r}}]}{2l_{\text{r}} + s_{\text{r}}} \\
&= \frac{2 \cdot [0.003(0.059 - 0.028) - 2 \cdot 0.0135 \cdot 0.0008]}{2 \cdot 0.0135 + 0.003} = 0.0048 \text{ m.}
\end{aligned}
$$

7.5.2 Determination of Resistance Coefficient

The resistance coefficient ζ_0 is calculated using Eq. [3.2] that is valid in the case of staggered arrangement of tubes in the bundle. We now determine the exponent n (Eq. [3.3]), the factor C_{r} (Eq. [3.4]), and the factor C_z' (Eqs. [3.13] and [3.15]), entering into Eq. [3.2]:

$$
\begin{aligned}
n &= 0.17 \left(\frac{A_{\text{rdc}}}{F} \right)^{0.25} \left(\frac{S_1}{S_2} \right)^{0.57} \exp\left(-0.36 \frac{S_1}{S_2} \right) \\
&= 0.17 \cdot (53.95)^{0.25} \cdot (1.16)^{0.57} \cdot \exp(-0.36 \cdot 1.16) = 0.330;
\end{aligned}
$$

$$
\begin{aligned}
C_{\text{r}} &= 2.8 \left(\frac{A_{\text{rdc}}}{F} \right)^{0.53} \left(\frac{S_1}{S_2} \right)^{1.30} \exp\left(-0.90 \frac{S_1}{S_2} \right) \\
&= 2.8 \cdot (53.95)^{0.53} \cdot (1.16)^{1.30} \cdot \exp(-0.90 \cdot 1.16) = 9.897;
\end{aligned}
$$

$C_z' = 1.0$, since $z_2 = 18$, that is, $z_2 \geq 6$.

According to Eq. [3.2], the resistance coefficient of the bundle ζ_0 is determined:

$$
\zeta_o = C_z' \cdot C_{\text{r}} \cdot \left(\frac{u_{\text{g}} \cdot d_{\text{eq}}}{\nu_{\text{g}}} \right)^{-n} = 1.0 \cdot 9.897 \cdot \left(\frac{10.84 \cdot 0.0048}{0.9945 \times 10^{-5}} \right)^{-0.33} = 0.587.
$$

7.5.3 Calculation of Aerodynamic Resistance of WAHE

Using Eq. [3.1], the aerodynamic resistance of WAHE is calculated:

$$\Delta H = c_{op} \cdot \zeta_o \cdot z_2 \cdot \frac{\rho_g \cdot u_g^2}{2} = 1.1 \cdot 0.587 \cdot 18 \cdot \frac{2.049 \cdot 10.84^2}{2} = 1400 \text{ Pa.}$$

7.6 CALCULATION OF HYDRAULIC RESISTANCE OF WAHE

7.6.1 Geometric Characteristics Needed for Calculations (Fig. 7.1)

Water moves in 18 coils connected in parallel. The calculation is performed for a single coil.

The total length of the coil L_0 is composed of the lengths of heated L_h and unheated L_{un} segments:

$$L_0 = L_h + L_{un}.$$

In the considered case, the length of the heated part of the coil L_h is made up of the lengths of straight finned segments within the confines of the gas conduit $L_{c \cdot cr \cdot s}$, arrangement of the bundle is determined:

$$n_{str} = \frac{z_2}{2}$$

then,

$$L_h = L_{c \cdot cr \cdot s} \cdot n_{str} = L_{c \cdot cr \cdot s} \cdot \frac{z_2}{2} = 0.5 \cdot \frac{18}{2} = 4.5 \text{ m.}$$

The length of the unheated part L_{un} in the considered case is composed of the lengths of the segment of withdrawal from the supply collector L_{wdr} and the segment of feed to the intake collector L_{fd} (Fig. 7.1), and the lengths of bends L_{bnd}:

$$L_{un} = L_{wdr} + L_{bnd} \cdot n_{bnd} + L_{fd}.$$

The number of the coil bends with a staggered arrangement of tubes in the bundle is shown:

$$n_{bnd} = \frac{z_2}{2} - 1 = \frac{18}{2} - 1 = 8.$$

The length of a single bend is as follows:

$$L_{bnd} = \pi R.$$

The bend radius R in the case of staggered arrangement is equal to the longitudinal tube spacing:

$$R = S_2 = 0.051 \text{ m};$$

the lengths of the feed and withdrawal segments are taken to be this:

$$L_{wdr} = L_{fd} = 0.3 \text{ m};$$

thus,

$$L_{uh} = 0.3 + \pi \cdot 0.051 \cdot 8 + 0.3 = 1.88 \text{ m}$$
$$L_0 = L_h + L_{uh} = 4.5 + 1.88 = 6.38 \text{ m}.$$

In WAHE, the medium is fed and withdrawn through horizontal collectors with the inside diameter $d_{col} = 0.081$ m and the wall thickness $\delta_{col} = 0.004$ m. The medium feed to the supply collector and the medium withdrawal from the intake collector are end–type, with a full cross–section to one side, ie, a Π–shaped scheme of the medium motion is implemented in WAHE.

Since a monophase medium (water) moves in WAHE, its hydraulic resistance is calculated using equations for monophase media.

7.6.2 Calculation of Hydraulic Resistance of Coil ΔP_{coil}

1. Coefficient of hydraulic friction resistance ζ_{fr}

According to Section 4.4.2 of Chapter 4, to determine ζ_{fr}, it is necessary to select a value of the absolute roughness of tubes $\ni$ from Table 4.1. Since the interior of bimetallic tubes is made of carbon steel (Section 7.3), then $\ni = 8.0 \times 10^{-5}$ m.

The relative roughness of tubes is determined:

$$\frac{\ni}{d_{in}} = \frac{8 \times 10^{-5}}{2 \times 10^{-2}}; \quad \frac{d_{in}}{\ni} = \frac{2 \times 10^{-2}}{8 \times 10^{-5}} = 250;$$

the boundary of the self–similar region of the resistance law at such relative roughness corresponds to the following boundary value of the Reynolds number:

$$Re_f^{bound} = 560 \cdot \frac{d_{in}}{\ni} = 560 \cdot 250 = 140 \times 10^3.$$

Since the actual value of the Reynolds number, Re_f (Section 7.4.4) is smaller than the boundary value,

$$Re_f = 13,358 < \mathbf{Re}_f^{bound} = 140 \times 10^3,$$

ζ_{fr} should be determined from Fig. 4.1:

$$\text{at } Re_f = 13,358 \text{ and } \frac{d_{in}}{\ni} = 250: \zeta_{fr} = 0.0325.$$

2. Resistance coefficient of the entrance to the heated tube ζ_{ent}
   In the considered case, there is an end feed of the medium to the supply collector (Fig. 4.3, 2) and therefore, according to Section 4.5.2 and Table 4.2,

$$\frac{d_{in}}{d_{col}} = \frac{0.02}{0.081} = 0.247 > 0.1 \text{ and } \zeta_{ent} = 0.7.$$

3. Resistance coefficient of the exit from the heated tube to the intake collector ζ_{ex}
   According to Section 4.5.3 of Chapter 4 and Table 4.3,

$$\zeta_{ex} = 1.1.$$

4. Resistance coefficients of turns (bends) ζ_{bnd}
   The resistance coefficient of a bend is determined as a function of the angle of turn in the bend φ and of the relative radius of the bend R/d_{in} according to Section 4.5.4 of Chapter 4, Table 4.4, or Fig. 4.4.
   In all bends, the flow turns through the same angle $\varphi = 180$ degree. According to Section 7.6.1,

$$R = 0.051 \text{ m}, \quad \frac{R}{d_{in}} = \frac{0.051}{0.020} = 2.55;$$

then, according to Table 4.4,

$$\zeta_{bnd} = 0.356.$$

The total resistance coefficient of the coil bends is found:

$$\sum \zeta_{bnd} = n_{bnd} \cdot \zeta_{bnd} = 8 \cdot 0.356 = 2.848.$$

5. The hydraulic resistance of the coil
   Since the water density is practically invariable along the entire length of the coil and its velocities in the characteristic cross-sections are equal,

the hydraulic resistance of the tube part of WAHE can be determined as follows:

$$\Delta P_{coil} = \left(\frac{\zeta_{fr}}{d_{in}} \cdot L_0 + \zeta_{ent} + n_{bnd} \cdot \zeta_{bnd} + \zeta_{ex} \right) \frac{\rho_f \cdot u_f^2}{2}$$

$$= \left(\frac{0.0325}{0.02} \cdot 6.38 + 0.7 + 8 \cdot 0.356 + 1.1 \right) \frac{1000 \cdot (0.53)^2}{2}$$

$$= 2109 \text{ Pa}.$$

7.6.3 Total Pressure Loss in Supply and Intake Collectors of WAHE $\overline{\Delta P}_{col}$

In accordance with Section 4.6.4 of Chapter 4, for determining $\overline{\Delta P}_{col}$, it is necessary to know maximum variations in the static pressure in the supply $\overline{\Delta P}_{col}^{sup}$ and intake $\overline{\Delta P}_{col}^{int}$ collectors, which are calculated according to Section 4.6.3 of Chapter 4.

The free areas of the collectors are shown:

$$f_{col}^{sup} = f_{col}^{int} = f_{col} = \frac{\pi \cdot d_{col}^2}{4} = \frac{\pi \cdot (0.081)^2}{4} = 5.153 \times 10^{-3} \text{ m}^2.$$

Maximum velocities of water in the supply and intake collectors are practically identical, since its density in the considered temperature range is constant:

$$u_{f_{max}}^{sup} = u_{f_{max}}^{int} = u_{f_{max}} = \frac{G_f \cdot v_f}{f_{col}} = \frac{3.0 \cdot 0.001}{5.153 \times 10^{-3}} = 0.58 \text{ m/s}.$$

From Table 4.6, the coefficient B is determined that takes account of the pressure loss in the collector:
- for the supply collector (with an end feed with a full cross section), $B_{sup} = 0.8$
- for the intake collector (with an end withdrawal), $B_{int} = 2.0$

The maximum variation in the static pressure in the supply collector of WAHE is determined:

$$\Delta P_{col}^{sup} = B_{sup} \cdot \frac{\rho_f u_{f_{max}}^2}{2} = 0.8 \cdot \frac{1000 \cdot (0.58)^2}{2} = 135 \text{ Pa}.$$

The maximum variation in the static pressure in the intake collector is as follows:

$$\Delta P_{col}^{int} = B_{int} \cdot \frac{\rho_f u_{f\,max}^2}{2} = 2.0 \cdot \frac{1000 \cdot (0.58)^2}{2} = 336 \text{ Pa}.$$

According to Section 4.6.4 of Chapter 4, $\overline{\Delta P}_{col}$ with a Π-shaped scheme of the medium motion in the element is determined from Eq. [4.16]:

$$\overline{\Delta P}_{col} = \frac{2}{3}\left(\Delta P_{col}^{int} - \Delta P_{col}^{sup}\right) = \frac{2}{3}(336 - 135) = 134 \text{ Pa}.$$

7.6.4 Total Hydraulic Resistance of WAHE ΔP_{el}

According to Section 4.2 of Chapter 4, the total hydraulic resistance of the tube element is made up of the hydraulic resistance of the tube part of the element (the coil) ΔP_{coil} and the total pressure loss in its collectors $\overline{\Delta P}_{col}$:

$$\Delta P_{el} = \Delta P_{coil} + \overline{\Delta P}_{col} = 2109 + 134 = 2243 \text{ Pa}.$$

7.6.5 Conclusions Based on Calculations

Comparing the calculated results with the calculation assignment, it can be concluded that the WAHE structure, adopted in the first version, does not comply with limitations on its aerodynamic resistance:

$$\Delta H = 1400 \text{ Pa} > \Delta H_{ad} = 1000 \text{ Pa}.$$

We now perform the calculation for the second version of the WAHE structure, differing from the first by the arrangement of tubes in the bundle. In lieu of the maximally dense "equilateral" arrangement, a staggered arrangement is assumed that corresponds to a maximum value of the co-efficient of convective heat transfer from the outside h_c at the specified fin coefficient of tubes $\Psi_r = 14.6$. Using recommendations in Section 2.4.5 of Chapter 2, the ratio of transverse and lateral tube spacings is determined that complies with this condition:

$$\left(\frac{S_1}{S_2}\right)_{max} = \frac{1.26}{\Psi_r} + 2 = \frac{1.26}{14.6} + 2 = 2.08.$$

Table 7.1 Calculation results

No.	Determined quantity	Source	Designation	Calculated results		Unit
				Ver. 1	Ver. 2	
1	2	3	4	5	6	7
Calculation of heat transfer						
1.	Transverse tube spacing	—	S_1	0.059	0.086	m
2.	Longitudinal tube spacing	—	S_2	0.051	0.041	m
3.	Diagonal tube spacing	—	S_2'	0.059	0.059	m
4.	Ratio of tube spacings	—	S_1/S_2	1.16	2.10	—
5.	Number of tubes in transverse row of bundle	—	z_1	9	6	Items
6.	Conventional diameter of finned tube	[2.22]	d_{cl}	0.0352	0.0352	m
7.	Parameter of tube bundle	[2.21]	φ_{cl}	1.0	2.13	—
8.	Flow area for air	[2.23] [2.24]	F	0.1216	0.1638	m²
9.	Calculated air velocity	[2.20]	u_g	10.84	8.04	m/s
10.	Number of tubes connected in parallel	[2.49]	z_{tcp}	18	12	—
11.	Flow area for water	[2.48]	F	5.66×10^{-3}	3.78×10^{-3}	m²
12.	Average water velocity in tubes	[2.47]	u_f	0.53	0.79	m/s
13.	Thermal efficiency	Section 2.6.6	ψ	0.95	0.95	—
14.	Thermal contact resistance	—	R_{cont}	1.89×10^{-4}	1.89×10^{-4}	m² K/W
15.	Ratio of outside and inside surface areas of finned tubes	—	$\frac{A}{A_{in}}$	20.45	20.45	—
16.	Shape factor of bundle	[2.31]	X	−0.926	0	—
17.	Exponent in Reynolds number	[2.29]	n	0.715	0.773	—
18.	Factor in equation for h_c	[2.30]	C_q	0.0725	0.0472	—

19.	Factor taking account of effect of number of transverse tube rows in bundle on heat transfer	[2.37]	C_z	1.0	1.0	—
20.	Convective heat transfer coefficient from outside	[2.28]	h_c	123.1	115.8	W/m² K
21.	Average fin temperature	[2.16]	T_r	28.0	28.0	°C
22.	Thermal conductivity of fin	—	k_r	225	225	W/m K
23.	Fin parameter	[2.13]	β	36.98	35.87	m⁻¹
24.	Conventional fin height	[2.14]	l'_r	0.0162	0.0162	m
25.	Theoretical efficiency of the fin	[2.11]	E	0.895	0.901	—
26.	Correction factor to theoretical efficiency of fin E	[2.15]	ψ_E	0.985	0.985	—
27.	Correction factor to theoretical efficiency of fin E, taking account of fin widening toward base	Fig. 2.1	μ_r	1.0	1.0	—
28.	Reduced heat transfer coefficient	[2.10]	h_{1rdc}	109.2	103.4	W/m² K
29.	Reynolds number for internal heat-transfer medium	—	$\mathbf{Re}_f$	13,538	20,179	—
30.	Parameter in equation for h_2	[2.44]	K	1.066	1.045	—
31.	Resistance coefficient in equation for h_2	[2.45]	ζ	0.0289	0.0261	—
32.	Correction for effect of temperature dependence of physical properties of heat-transfer medium on h_2	[2.50]	C_{tem}	1.021	1.021	—
33.	Heat transfer coefficient to internal medium	[2.43]	h_2	3120	4363	W/m² K
34.	Overall heat transfer coefficient	[2.6]	U	48.51	52.12	W/m² K
35.	Average temperature difference	[2.53]	ΔT	31.8	31.8	°C
36.	Calculated area of heat-transfer surface	[2.1]	A	97.43	90.69	m²
37.	Total number of tubes in WAHE	—	z	162	144	—
38.	Number of transverse tube rows	—	z_2	18	24	—

Continued

Table 7.1 Calculation results—cont'd

No.	Determined quantity	Source	Designation	Calculated results		Unit
				Ver. 1	Ver. 2	
1	2	3	4	5	6	7
39.	Depth of gas conduit	—	c	0.870	0.984	m
40.	Actual length of finned tubes of WAHE	—	L_r^{ac}	81	72	m
41.	Actual area of heat transfer surface	—	A^{ac}	104.0	92.4	m^2
Calculation of aerodynamic resistance						
42.	Reduced length of developed surface	[3.9]	$\frac{A_{coal}}{F}$	53.95	25.28	—
43.	Equivalent diameter of most contracted cross-section of bundle	[3.10] [3.11]	d_{eq}	0.0048	0.0095	m
44.	Exponent on Reynolds number, Re_{eq}	[3.3]	n	0.330	0.273	—
45.	Factor in equation for resistance coefficient	[3.4]	C_r	9.897	6.147	—
46.	Correction factor taking into account the small number of rows in bundle	[3.15]	C_z'	1.0	1.0	—
47.	Resistance coefficient of single tube row	[3.2]	ζ_0	0.587	0.533	—
48.	Aerodynamic resistance of WAHE	[3.1]	ΔH	1400	932	Pa
Calculation of hydraulic resistance						
49.	Length of heated segments of coil	—	L_h	4.5	6.0	m
50.	Number of coil bends	—	n_{bnd}	8	11	—
51.	Length of unheated segments of coil	—	L_{un}	1.88	2.02	m
52.	Total length of coil	—	L_0	6.38	8.02	m
53.	Absolute roughness of tubes	Section 4.4.2	$\ni$	8×10^{-5}	8×10^{-5}	m

No.	Description	Reference	Symbol			Unit
54.	Ratio $d_{in}/∋$	—	$d_{in}/∋$	250	250	—
55.	Friction resistance coefficient	Fig. 4.1	ζ_{fr}	0.0325	0.0310	—
56.	Inside diameter of collectors	—	d_{col}	0.081	0.081	m
57.	Ratio d_{in}/d_{col}	—	d_{in}/d_{col}	0.247	0.247	—
58.	Resistance coefficient of entrance to heated tube	Table 4.2	ζ_{ent}	0.7	0.7	—
59.	Resistance coefficient of exit from heated tube to intake collector	Table 4.3	ζ_{ex}	1.1	1.1	—
60.	Bend radius	—	R	0.051	0.041	m
61.	Ratio R/d_{in}	—	R/d_{in}	2.55	2.05	—
62.	Angle of turn of flow in bend	—	φ	180	180	degree
63.	Resistance coefficient of bend	Table 4.4	ζ_{bnd}	0.356	0.396	—
64.	Total resistance coefficient of coil bends	—	$n_{bnd}\cdot\zeta_{bnd}$	2.848	4.356	—
65.	Hydraulic resistance of coil	—	ΔP_{coil}	2109	5800	Pa
66.	Free area of collector	—	f_{col}	5.153×10^{-3}	5.153×10^{-3}	m²
67.	Maximum medium velocity in collector	—	u_{fmax}	0.58	0.58	m/s
68.	Factor taking into account pressure losses in distributing collector	Table 4.6	B_{sup}	0.8	0.8	—
69.	Factor taking into account pressure losses in intake collector	Table 4.6	B_{int}	2.0	2.0	—
70.	Maximum variation in static pressure in distributing collector	[4.15]	ΔP_{col}^{sup}	135	135	Pa
71.	Max. variation in static pressure in intake collector	[4.15]	ΔP_{col}^{int}	336	336	Pa
72.	Total pressure loss in distributing and intake collectors of WAHE	[4.16]	$\overline{\Delta P_{col}}$	134	134	Pa
73.	Total hydraulic resistance of WAHE	[4.5]	ΔP_{el}	2243	5934	Pa

Leaving the dimensions of the gas conduit unchanged ($a = 0.56$ m and $b = 0.50$ m), we determine the absolute values of tube spacings:

$$S_1 = \frac{a}{z_1 + 0.5} = \frac{0.560}{6 + 0.5} = 0.086 \text{ m};$$

$$S_2 = \frac{S_1}{(S_1/S_2)_{max}} = \frac{0.086}{2.08} = 0.041 \text{ m}.$$

Here, $z_1 = 6$ is the maximum number of finned tubes, fitting into the width $a = 0.56$ m with the ratio of tube spacings $(S_1/S_2) = 2.08$ and a sufficient technological clearance between tubes. In the considered case, the minimum clearance between tubes is determined by the finning diameter D and the diagonal tube spacing S_2':

$$S_2' = \sqrt{\frac{1}{4}S_1^2 + S_2^2} = \sqrt{\frac{1}{4}(0.086)^2 + (0.041)^2} = 0.0594 \text{ m} > D$$
$$= 0.055 \text{ m}.$$

The results, calculated for the second WAHE version in Sections 7.3.2–7.6.4, as compared with those for the first WAHE version, are presented in Table 7.1.

Comparison of the results, calculated for two WAHE versions, indicates that the selection of the rational ratio of the tube spacings in the heat exchanger made it possible to reduce by 50% the aerodynamic resistance of the heat exchanger, which mainly determines the energy consumption during its operation, and to keep within the limits of the imposed restrictions ($\Delta H < 0.001$ MPa). Furthermore, it provided a 12.5% decrease in the mass of the most expensive part of the heat exchanger, ie, the bundle of bimetallic finned tubes.

APPENDIX A

Units Conversion

Primary or Fundamental Dimensions and their Units in SI (International System)

No.	Dimension	Unit
1	Length	meter (m)
2	Mass	kilogram (kg)
3	Time	second (s)
4	Temperature	kelvin (K)
5	Electric current	ampere (A)
6	Amount of light	candela (cd)
7	Amount of matter	mole (mol)

Standard Prefixes in SI Units

Multiple	Prefix
10^{12}	tera (T)
10^{9}	giga (G)
10^{6}	mega (M)
10^{3}	kilo (k)
10^{2}	hector (h)
10^{1}	deka (d)
10^{-1}	deci (d)
10^{-2}	centi (c)
10^{-3}	milli (m)
10^{-6}	micro (μ)
10^{-9}	nano (n)
10^{-12}	pico (p)

Units Conversion
Area
$1 \text{ m}^2 = 10.76391 \text{ ft}^2$.

$1 \text{ ft}^2 = 0.092903 \text{ m}^2$; $1 \text{ in}^2 = 6.4516 \text{ cm}^2 = 645.16 \text{ mm}^2$; $1 \text{ mil}^2 = 6.452 \times 10^{-4} \text{ mm}^2$; 1 circular mm $= 0.7853982 \text{ mm}^2$; 1 circular (circ.) in $= 506.707479 \text{ mm}^2$; 1 circ. mil $= 5.0671 \times 10^{-4} \text{ mm}^2$; $1 \text{ yd}^2 = 0.8361 \text{ m}^2$; 1 acre $= 4840 \text{ yd}^2 = 4046.86 \text{ m}^2$; $1 \text{ mi}^2 = 2.59 \text{ km}^2$.

Density

1 kg/m^3 = 6.24280 × 10^{-2} lb/ft^3.
1 lb/ft^3 = 16.0185 kg/m^3; 1 slug/ft^3 = 515.38 kg/m^3.

Electrical Resistivity Specific

1 ohm circ. mil/ft = 1.6624261 × 10^{-9} ohm m; 1 ohm in = 0.0254 ohm m.
Sometimes, instead of "Ohm," a symbol is used, "Ω."

Energy, Work, and Heat Amount

Units	J[a]	erg	kg$_f$m	kcal	kWh
1J=1Ws	1.0	10^7	0.101972	2.38846 × 10^{-4}	2.7778 × 10^{-7}
1 erg	10^{-7}	1.0	1.01972 × 10^{-8}	2.38846 × 10^{-11}	2.7778 × 10^{-14}
1 kg$_f$m	9.80665	9.80665 × 10^7	1.0	2.34228 × 10^{-3}	2.724 × 10^{-6}
1 kcal	4.1868 × 10^3	4.1868 × 10^{10}	4.26935 × 10^2	1.0	1.163 × 10^{-3}
1 kWh	3.6 × 10^6	3.6 × 10^{13}	3.67098 × 10^5	8.59845 × 10^2	1.0

[a]Units based on names of scientists (researchers, etc.) should be capitalized.

1 horse power per hour (h.p. h) = 2647.8 kJ; 1 eV (eV) = 1.602 × 10^{-19} J;
1 erg = 1 dyne (dyn) cm = 0.1 × 10^{-6} J, 1 L (L, l or ℓ) atm = 101.33 J.

1 British thermal unit (B.t.u. or BTU) (thermal) = 1.05435 kJ;
1 B.t.u. = 1.055056 kJ; 1 B.t.u. (mean) = 1.05587 kJ.

1 calorie (cal) (thermal) = 4.184 J; 1 cal = 4.1868 J; 1 cal (at 15°C) = 4.1858 kJ; 1 cal (at 20°C) = 4.1819 J; 1 cal (mean) = 4.19002 J; 1 Calorie (Cal) (food) = 4.1868 kJ.

1 lb$_f$ ft = 1.355817 J; 1 poundal (pdl) ft = 0.04214 J; 1 therm = 105,506 kJ;
1 pound centigrade unit (pcu) = 1.8 B.t.u. = 1.8978 kJ.

Specific Enthalpy

1 kcal/kg = 1 cal/g = 4.1868 kJ/kg; 1 B.t.u./lb = 2.326 kJ/kg; 1 centigrade heat unit (chu)/lb = 4.1868 kJ/kg; 1 ft lb$_f$/lb = 0.00299 kJ/kg.

Flowrate Volumetric (or Volume Flowrate)

1 gallon (US)/min (gpm) = 0.06309 dm^3/s; 1 ft^3/min (cfm) = 471.95 dm^3/s.

Force

1 N = 10^5 dyn = 0.101972 kg$_f$; 1 kg$_f$ = 9.80665 N.

1 lb$_f$ = 4.448222 N; 1 cental = 100 lb$_f$ = 444.822 N; 1 kip = 1000 lb$_f$ = 4448.222 N; 1 poundal (pdl) = 1 lb$_f$ ft/s^2 = 0.138255 N; 1 grain = 0.6355 × 10^{-3} N; 1 stone = 62.2751 N.

1 lb$_f$ = 0.01 cental = 0.001 kip = 32.174 pdl = 7000.00 grain = 16 oz$_f$.

Heat Flux

1 W/m^2 = 0.8598 kcal/h m^2 = 7.988 × 10^{-2} kcal/h ft^2 = 0.31791 B.t.u./h ft^2 = 0.17611 chu/h ft^2.

1 kcal/h m^2 = 1.163 W/m^2; 1 cal/s cm^2 = 41,868 W/m^2.

1 kcal/h ft^2 = 12.5184 W/m^2; 1 B.t.u./h ft^2 = 3.154 W/m^2;
1 chu/h ft^2 = 5.6783 W/m^2.

Heat Flux Volumetric
1 B.t.u./h ft^3 = 10.35 W/m^3.

Heat Transfer Coefficient
1 B.t.u./h ft$^{2\circ}$F = 5.6782 W/m^2 K.

Heat Transfer Rate and Power

Units	W	kcal	kg$_f$ m/s	h.p. (metric)
1 W = 1 J/s	1.0	0.86	0.102	1.36 × 10^{-3}
1 kcal/h	1.163	1.0	0.118	1.58 × 10^{-3}
1 kg$_f$ m/s	9.81	8.44	1.0	1.33 × 10^{-2}
1 h.p. (metric)	735.5	633	75	1.0

1 W = 10^7 erg/s; 1 m^3 atm/h = 28.146 W; 1 h.p. (el.) = 746 W.

1 horse power (h.p.) (UK) = 745.7 W; 1 B.t.u./h = 0.29307 W;
1 lb$_f$ ft/min = 0.022597 W; 1 lb$_f$ ft/s = 1.35582 W; 1 cheval-vapor = 1 h.p.
(metric) = 735.5 W; 1 chu/h = 0.52753 W.

Length
1 m = 3.2808 ft.

1 yard (yd) = 0.9144 m = 3 ft (3′); 1 ft = 12 inch (in) (12″).

1 ft = 0.3048 m; 1 in (1″) = 25.4 mm; 1 mil = 0.001 in = 0.0254 mm;
1 μm = 10^{-6} m; 1 mile (mi) = 5280 ft = 1609.344 m; 1 nautical mile
(nmi) = 6076.1 ft = 1852 m.

Mass
1 pound (lb) = 16 ounces (oz) = 453.59237 g; 1 oz = 28.3495 g; 1 slug =
32.174 lb = 14.594 kg; 1 short ton (or tonne) (t) (US 2000 lb) = 0.9072 metric
ton; 1 long ton (imperial ton, UK 2240 lb) = 1.016 metric ton.

Pressure

Units	Pa	bar	kg$_f$/cm^2	atm (phys.[a])	mm Hg	mm H$_2$O
1 Pa = 1 N/m^2 =	1	10^{-5}	1.02 × 10^{-5}	0.987 × 10^{-5}	7.5024 × 10^{-3}	0.10197
1 bar =	10^5	1.0	1.02	0.98692	7.5024 × 10^2	1.02 × 10^4
1 kg$_f$/cm^2 = 1 at (tech.[a]) =	9.80665 × 10^4	0.980665	1.0	0.9678	735.56	10^4

Continued

Units	Pa	bar	kg_f/cm^2	atm (phys.[a])	mm Hg	mm H_2O
1 atm (phys.) =	1.01325×10^5	1.01325	1.0332	1.0	760.0	1.0332×10^4
1 mm Hg = 1 torr =	133.322	1.33×10^{-3}	1.36×10^{-3}	1.316×10^{-3}	1.0	13.595
1 mm H_2O = 1 kg_f/m^2 =	9.80665	9.80665×10^{-5}	10^{-4}	9.678×10^{-5}	7.356×10^{-2}	1.0

[a]*Phys.*, physical; *tech.*, technical.

1 bar $= 10^6$ dyn/cm^2 $= 14.5038$ lb$_f$/in^2 (psi) $= 2088.543$ lb$_f$/ft^2 $= 29.530$ inches of Hg $= 401.463$ inches of water $= 1.4504 \times 10^{-2}$ kip/in $= 69,053.14$ poundal/ft^2.

1 lb$_f$/in^2 (pounds per square inch (psia)) $= 6894.76$ Pa; 1 lb$_f$/ft$^2 = 47.88$ Pa; 1 inch of Hg $= 3.3864$ kPa; 1 inch of water $= 249.1$ Pa; 1 kip/in $= 6894.76$ kPa.

Specific Heat

1 B.t.u./lb Ra $= 4186.9$ J/kg K; 1 ft lb$_f$/slug Ra $= 0.16723$ J/kg K.

Temperature Scales

Scale	K (*T*)	°C (*t*)	Ra (*T*)	°F (*t*)	°R (*t*)
Kelvin, K =	1.0	$t + 273.15$	$5/9T$	$5/9t + 255.37$	$5/4t + 273.15$
Celsius, °C =	$T - 273.15$	$1.0t$	$5/9t - 273.15$	$5/9(t - 32)$	$1.25t$
Rankine, Ra =	$1.8T$	$1.8(t + 273.15)$	$1.0T$	$t + 459.67$	$1.8(1.25t + 273.15)$
Fahrenheit, °F =	$1.8T - 459.67$	$1.8t + 32$	$T - 459.67$	$1.0t$	$2.25t + 32$
Réaumur, °R =	$0.8(T - 273.15)$	$0.8t$	$0.8(5/9T - 273.15)$	$(t - 32)4/9$	$1.0t$

The Rankine temperature scale (widely used in the United States, Canada, and other countries) is the absolute scale, 0 Ra = 0 K, at the same time 1 Ra = 1°F; sometimes, degrees of Rankine have a symbol "R," for example, 0 R; the same symbol is related to degrees of Réaumur; however, the sign of degree should be used in this case, ie, 0°R.

The temperature scale of Fahrenheit is the practical scale (also widely used in the United States, Canada, and other countries), 32°F = 0°C and 212°F = 100°C.

The Réaumur temperature scale is the practical scale, 0°R = 0°C, but 80°R = 100°C (currently, this scale is used rarely).

Temperature Difference

$\Delta T = 1$ K $= 1°C = (9/5)$ Ra $= (9/5)°F$.

Degrees of absolute scales are more preferable to be used for the temperature difference.

Thermal Conductivity

1 W/m K = 1 W/m°C = 0.8598 kcal/h m°C = 2.3885 cal/s cm°C = 0.5778 B.t.u./h ft°F = 0.5778 chu/h ft°C.

1 kcal/h m°C = 1.163 W/m K; 1 cal/s cm°C = 418.68 W/m K; 1 B.t.u./h ft°F = 1 chu/h ft°C = 1.7307 W/m K; 1 B.t.u. in/h°F ft^2 = 0.144 W/m K.

Viscosity Dynamic

Units	Pa s	kg/m s	kg$_f$ s/m^2	P (poise)
1 Pa s = 1 N s/m^2 =	1.0	1.0	0.101972	10.0
1 kg/m s =	1.0	1.0	0.101972	10.0
1 kg$_f$ s/m^2 =	9.80665	9.80665	1.0	98.0665
P (poise) =	0.1	0.1	0.101972 × 10^{-1}	1.0

1 Pa s = 0.671969 lb/ft s = 2419.088 lb/ft h = 2.08855 × 10^{-2} lb$_f$ s/ft^2.

1 lb/ft s = 1.4882 Pa s; 1 lb/ft h = 0.41338 × 10^{-3} Pa s; 1 lb$_f$ s/ft^2 = 1 slug/ft s = 47.8803 Pa s.

Viscosity Kinematic

1 m^2/s = 1 × 10^4 St (stokes) = 10.7639 ft^2/s = 38,750.0775 ft^2/h = 91,440.0 L (L, l or ℓ)/in h;

1 St = 1 × 10^{-4} m^2/s.

Volume

1 m^3 = 35.3147 ft^3; 1 ℓ = 1 dm^3 = 0.001 × m^3.

1 ft^3 = 0.028317 × m^3; 1 in^3 = 16.3871 cm^3; 1 gallon liquid US = 3.7854 ℓ and 1 gallon UK = 4.54609 ℓ; 1 fluid oz (UK) = 28.413 mℓ; 1 fl oz (US) = 29.574 mℓ; 1 pint (pt) (UK) = 0.5683 dm^3.

Notes: at = atmosphere (technical), atm = atmosphere (physical), B.t.u. (Btu or BTU) = British thermal unit; cal = calorie, cc = cubic centimeter, chu = centigrade heat unit, 1 circular mil = area of circle with diameter of 1 mil, dyn = dyne, eV = electronvolt, f = force, fl = fluid, ft = foot or feet, Hg = mercury, in = inch, J = joule, h = hour, h.p. (hp) = horse power, ℓ (L or l) = liter (liter), lb (from Latin: libra) = pound or lb$_m$ (lbm) = pound of mass, lb$_f$ (lbf) = pound of force, m = meter (meter), mi = mile; mil = unit of length equal to one-thousandth of an inch, min = minute, μm = micrometer; N = newton, nmi = nautical mile, oz = ounce,

P = poise, Pa = pascal, pcu = pound centigrade unit, pdl = poundal, psi = pounds per square inch, psia = psi absolute, psig = psi gauge, s = second, St = stoke, UK = United Kingdom (ie, British unit), US = United States, yd = yard, W = watt.

Some Physical Constants and Definitions

Normal acceleration due to gravity, g_n = 9.80665 m/s^2 = 32.174 ft/s^2;

Universal gas constant, R = 8.31451 J/mol K = 8.31447 kPa m^3/kmol K = 0.0831447 bar m^3/kmol K = 82.05 L atm/kmol K = 1.9858 Btu/lbmol Ra = 1545.37 ft lb$_f$/lbmol Ra = 10.73 psia ft^3/lbmol Ra.

Normal conditions are physical conditions at a pressure of p = 101,325 Pa = 1.01325 bar = 14.696 psi = 760 mm Hg (normal atmosphere) = 29.9213 in Hg = 10.3323 m H$_2$O and a temperature of t = 273.15 K = 0°C at which a molar gas volume is V_o = 2.24141 $\times$ 10^{-2} m^3/mol.

APPENDIX B

Recommended Thermophysical Properties Software for Gases and Liquids

Thermophysical properties of air, water, and other gases and fluids at different pressures and temperatures can be calculated using the NIST software (2010) (http://www.nist.gov/srd/nist23.cfm).

Version 9.1 includes 121 pure fluids, 5 pseudo-pure fluids (such as air), and mixtures with up to 20 components:
- *the typical natural gas constituents: methane, ethane, propane, butane, isobutane, pentane, isopentane, hexane, isohexane, heptane, octane, nonane, decane, undecane, dodecane, carbon dioxide, carbon monoxide, hydrogen, nitrogen, and water*
- *the hydrocarbons acetone, benzene, butene, cis-butene, cyclohexane, cyclopentane, cyclopropane, ethylene, isobutene, isooctane, methylcyclohexane, propylcyclohexane, neopentane, propyne, trans-butene, and toluene*
- *the HFCs R23, R32, R41, R125, R134a, R143a, R152a, R161, R227ea, R236ea, R236fa, R245ca, R245fa, R365mfc, R1233zd(E), R1234yf, and R1234ze(E)*
- *the refrigerant ethers RE143a, RE245cb2, RE245fa2, and RE347mcc (HFE-7000)*
- *the HCFCs R21, R22, R123, R124, R141b, and R142b*
- *the traditional CFCs R11, R12, R13, R113, R114, and R115*
- *the fluorocarbons R14, R116, R218, R1216, C4F10, C5F12, and RC318*
- *the "natural" refrigerants ammonia, carbon dioxide, propane, isobutane, and propylene*
- *the main air constituents: nitrogen, oxygen, and argon*
- *the noble elements helium, argon, neon, krypton, and xenon*
- *the cryogens argon, carbon monoxide, deuterium, krypton, neon, nitrogen trifluoride, nitrogen, fluorine, helium, methane, oxygen, normal hydrogen, parahydrogen, and orthohydrogen*
- *water (as a pure fluid, or mixed with ammonia)*
- *miscellaneous substances including carbonyl sulfide, diethyl ether, dimethyl carbonate, dimethyl ether, ethanol, heavy water, hydrogen chloride, hydrogen sulfide, methanol, methyl chloride, nitrous oxide, Novec-649, sulfur hexafluoride, sulfur dioxide, and trifluoroiodomethane*

- *the xylenes m-xylene, o-xylene, p-xylene, and ethylbenzene*
- *the FAMES (fatty acid methyl esters, ie, biodiesel constituents) methyl oleate, methyl palmitate, methyl stearate, methyl linoleate, and methyl linolenate*
- *the siloxanes octamethylcyclotetrasiloxane, decamethylcyclopentasiloxane, dodecamethylcyclohexasiloxane, decamethyltetrasiloxane, dodeca-methylpentasiloxane, tetradecamethylhexasiloxane, octamethyltrisiloxane, and hexamethyldisiloxane*
- *79 predefined mixtures (such as R407C, R410A, and air); the user may define and store others.*

This statement is taken from: http://www.nist.gov/srd/nist23.cfm, the Website accessed on July 17, 2015.

The mini-REFPROP program is a sample version of the full REFPROP program (located at www.nist.gov/srd/nist23.cfm) and is meant for use as a teaching tool in the introduction of thermodynamics to students. It contains a limited number of pure fluids (water, CO₂, R134a, nitrogen, methane, propane, hydrogen, and dodecane) and also allows mixture calculations of nitrogen with methane for teaching Vapor–Liquid Equilibrium (VLE). The program expires Aug. 31, 2016, at which point a new set of install files will be uploaded here.

This statement is taken from: http://www.boulder.nist.gov/div838/ theory/refprop/MINIREF/MINIREF.HTM, the Website accessed on July 17, 2015.

APPENDIX C

Tables of Properties of Substances

Table C.1 Thermophysical properties of dry air at 1 atm (101.325 kPa)

t	ρ	h	s	c_v	c_p	k	μ	Pr	$\beta_{th} \times 10^3$
°C	kg/m³	kJ/kg	kJ/kg K	kJ/kg K	kJ/kg K	W/m K	µPa s	—	1/K
−70	1.741	202.87	6.474	0.7163	1.0069	0.01848	13.55	0.738	4.963
−60	1.659	212.93	6.523	0.7163	1.0065	0.01930	14.10	0.736	4.726
−50	1.584	223.00	6.569	0.7163	1.0062	0.02011	14.65	0.733	4.510
−40	1.516	233.06	6.613	0.7164	1.0060	0.02090	15.19	0.731	4.313
−30	1.453	243.12	6.655	0.7165	1.0058	0.02169	15.72	0.729	4.133
−20	1.395	253.17	6.696	0.7166	1.0058	0.02247	16.24	0.727	3.968
−10	1.342	263.23	6.735	0.7169	1.0058	0.02324	16.75	0.725	3.815
0	1.293	273.29	6.772	0.7171	1.0059	0.02400	17.26	0.723	3.674
10	1.247	283.35	6.808	0.7175	1.0061	0.02475	17.76	0.722	3.543
20	1.204	293.41	6.843	0.7178	1.0064	0.02549	18.25	0.721	3.421
30	1.164	303.48	6.877	0.7183	1.0067	0.02622	18.73	0.719	3.307
40	1.127	313.55	6.910	0.7188	1.0072	0.02695	19.21	0.718	3.201
50	1.092	323.62	6.941	0.7194	1.0077	0.02766	19.68	0.717	3.101
60	1.059	333.70	6.972	0.7201	1.0083	0.02837	20.15	0.716	3.007
70	1.028	343.79	7.002	0.7208	1.0089	0.02907	20.60	0.715	2.919
80	0.999	353.88	7.031	0.7217	1.0097	0.02977	21.06	0.714	2.836
90	0.972	363.98	7.059	0.7226	1.0105	0.03046	21.51	0.714	2.758
100	0.946	374.09	7.087	0.7236	1.0115	0.03114	21.95	0.713	2.683
120	0.897	394.34	7.140	0.7258	1.0136	0.03249	22.82	0.712	2.546
140	0.854	414.64	7.190	0.7283	1.0160	0.03249	23.67	0.711	2.546
160	0.815	434.99	7.238	0.7311	1.0188	0.03511	24.50	0.711	2.310
180	0.779	455.39	7.284	0.7342	1.0219	0.03639	25.31	0.711	2.208
200	0.746	475.86	7.328	0.7376	1.0252	0.03765	26.11	0.711	2.115
250	0.674	527.35	7.432	0.7472	1.0347	0.04072	28.04	0.713	1.912
300	0.615	579.35	7.527	0.7580	1.0454	0.04369	29.89	0.715	1.745
400	0.524	685.04	7.697	0.7815	1.0688	0.04938	33.38	0.722	1.486
500	0.456	793.12	7.846	0.8054	1.0927	0.05479	36.63	0.731	1.293
600	0.404	903.54	7.980	0.8282	1.1154	0.05999	39.71	0.738	1.145
700	0.363	1016.10	8.103	0.8489	1.1361	0.06502	42.64	0.745	1.027
800	0.329	1130.70	8.215	0.8674	1.1545	0.06991	45.46	0.751	0.932
900	0.301	1247.00	8.318	0.8836	1.1708	0.07468	48.17	0.755	0.852
1000	0.277	1364.80	8.415	0.8978	1.1850	0.07936	50.80	0.758	0.785
1100	0.257	1483.90	8.505	0.9104	1.1975	0.08395	53.35	0.761	0.728
1200	0.240	1604.20	8.589	0.9214	1.2086	0.08848	55.85	0.763	0.679
1300	0.224	1725.60	8.669	0.9312	1.2184	0.09295	58.29	0.764	0.636
1400	0.211	1847.90	8.744	0.9400	1.2271	0.09738	60.69	0.765	0.598

Table C.2 Thermophysical properties of combustion products at 1 atm (101.325 kPa)
($P_{CO_2} = 0.13$, $P_{H_2O} = 0.11$, and $P_{N_2} = 0.76$)

t	ρ	h	c_p	k	μ	$a \times 10^6$	$v \times 10^6$	Pr
°C	kg/m³	kJ/kg	kJ/kg K	W/m K	µPa s	m²/s	m²/s	—
0	1.295	0	1.042	0.00228	15.8	16.9	12.2	0.722
100	0.950	105.5	1.068	0.00313	20.4	30.9	21.5	0.696
200	0.748	213.8	1.097	0.00401	24.5	48.9	32.8	0.670
300	0.617	324.7	1.122	0.00484	28.2	69.9	45.7	0.654
400	0.525	438.4	1.151	0.00570	31.7	94.3	60.4	0.640
500	0.457	555.2	1.185	0.00656	34.8	121.1	76.2	0.629
600	0.405	675.1	1.214	0.00742	37.9	150.9	93.6	0.620
700	0.368	797.8	1.239	0.00827	40.7	181.4	110.6	0.610
800	0.330	922.9	1.264	0.00915	43.4	219.4	131.5	0.600
900	0.301	1050.6	1.290	0.01000	45.9	257.5	152.5	0.592
1000	0.275	1180.4	1.306	0.01090	48.4	303.5	176.0	0.580
1100	0.257	1311.9	1.323	0.01175	50.7	345.6	197.3	0.571
1200	0.240	1445.0	1.340	0.01262	53.0	392.4	220.8	0.563
1300	0.230	1580.0	1.360	0.01351	55.0	431.9	239.1	0.554
1400	0.220	1717.0	1.380	0.01440	56.5	474.3	256.8	0.541
1500	0.210	1856.5	1.410	0.01530	58.6	516.7	279.1	0.540
1600	0.200	1999.0	1.440	0.01620	60.0	562.5	300.0	0.533
1700	0.195	2144.0	1.460	0.01705	64.5	598.9	330.8	0.552
1800	0.190	2291.0	1.480	0.01810	68.6	643.7	361.1	0.561
1900	0.185	2440.0	1.500	0.01895	72.2	682.9	390.3	0.572
2000	0.180	2591.0	1.520	0.01985	75.4	725.5	418.9	0.577

Table C.3 Thermophysical properties of water (liquid) on saturation line (based on temperature increments)

t	P_s	ρ	h	c_p	k	$\mu \times 10^6$	$\sigma \times 10^3$	$\beta_{th} \times 10^3$	Pr
°C	MPa	kg/m³	kJ/kg	J/kg K	W/m K	Pa s	N/m	1/K	—
5.0	0.0009	999.9	21.0	4205.5	0.571	1518.3	74.94	0.016	11.192
10.0	0.0012	999.7	42.0	4195.5	0.580	1306.0	74.22	0.088	9.447
15.0	0.0017	999.1	63.0	4188.8	0.589	1137.6	73.49	0.151	8.086
20.0	0.0023	998.2	83.9	4184.4	0.598	1001.6	72.74	0.207	7.004
25.0	0.0032	997.0	104.8	4181.6	0.607	890.1	71.97	0.257	6.130
30.0	0.0042	995.6	125.7	4180.1	0.615	797.4	71.19	0.303	5.416
35.0	0.0056	994.0	146.6	4179.5	0.623	719.3	70.40	0.346	4.823
40.0	0.0074	992.2	167.5	4179.6	0.631	653.0	69.60	0.385	4.328
45.0	0.0096	990.2	188.4	4180.4	0.637	596.1	68.78	0.423	3.910
50.0	0.0124	988.0	209.3	4181.5	0.644	546.8	67.94	0.458	3.553

Table C.3 Thermophysical properties of water (liquid) on saturation line (based on temperature increments)—cont'd

t	P_s	ρ	h	c_p	k	$\mu \times 10^6$	$\sigma \times 10^3$	$\beta_{th} \times 10^3$	Pr
°C	MPa	kg/m³	kJ/kg	J/kg K	W/m K	Pa s	N/m	1/K	—
55.0	0.0158	985.7	230.3	4183.1	0.649	504.0	67.10	0.491	3.247
60.0	0.0199	983.2	251.2	4185.1	0.654	466.4	66.24	0.523	2.983
65.0	0.0250	980.5	272.1	4187.5	0.659	433.2	65.37	0.554	2.753
70.0	0.0312	977.7	293.1	4190.2	0.663	403.9	64.48	0.584	2.552
75.0	0.0386	974.8	314.0	4193.3	0.667	377.7	63.58	0.613	2.376
80.0	0.0474	971.8	335.0	4196.9	0.670	354.3	62.67	0.641	2.220
85.0	0.0579	968.6	356.0	4200.8	0.673	333.3	61.75	0.669	2.081
90.0	0.0702	965.3	377.0	4205.3	0.675	314.4	60.82	0.697	1.958
95.0	0.0846	961.9	398.1	4210.2	0.677	297.3	59.87	0.724	1.848
100.0	0.1014	958.4	419.2	4215.7	0.679	281.7	58.91	0.751	1.749
105.0	0.1209	954.7	440.3	4221.7	0.681	267.6	57.94	0.777	1.660
110.0	0.1434	951.0	461.4	4228.3	0.682	254.7	56.96	0.804	1.580
115.0	0.1692	947.1	482.6	4235.6	0.683	242.9	55.97	0.831	1.507
120.0	0.1987	943.1	503.8	4243.5	0.683	232.1	54.97	0.858	1.441
130.0	0.2703	934.8	546.4	4261.5	0.684	212.9	52.93	0.912	1.327
140.0	0.3615	926.1	589.2	4282.6	0.683	196.5	50.86	0.968	1.232
150.0	0.4762	917.0	632.2	4307.1	0.682	182.5	48.74	1.027	1.152
160.0	0.6182	907.5	675.5	4335.4	0.680	170.2	46.59	1.088	1.085
170.0	0.7922	897.5	719.1	4367.8	0.677	159.6	44.41	1.153	1.029
180.0	1.0028	887.0	763.1	4405.0	0.673	150.1	42.19	1.222	0.982
190.0	1.2552	876.1	807.4	4447.4	0.669	141.8	39.95	1.297	0.943
200.0	1.5549	864.7	852.3	4495.8	0.663	134.3	37.68	1.379	0.910
210.0	1.9077	852.7	897.6	4551.2	0.657	127.6	35.38	1.469	0.884
220.0	2.3196	840.2	943.6	4614.6	0.650	121.5	33.07	1.569	0.863
230.0	2.7971	827.1	990.2	4687.6	0.641	116.0	30.74	1.682	0.848
240.0	3.3469	813.4	1037.6	4771.9	0.632	110.9	28.39	1.810	0.837
250.0	3.9762	798.9	1085.8	4870.1	0.621	106.1	26.04	1.958	0.832
260.0	4.6923	783.6	1135.0	4985.6	0.609	101.7	23.69	2.130	0.832
270.0	5.5030	767.5	1185.3	5123.0	0.596	97.5	21.34	2.334	0.838
280.0	6.4166	750.3	1236.9	5288.9	0.581	93.5	18.99	2.581	0.851
290.0	7.4418	731.9	1290.0	5493.1	0.565	89.7	16.66	2.886	0.872
300.0	8.5879	712.1	1345.0	5750.4	0.547	85.9	14.36	3.274	0.902
310.0	9.8651	690.7	1402.2	6084.8	0.529	82.2	12.09	3.784	0.946
320.0	11.2840	667.1	1462.2	6537.3	0.509	78.4	9.86	4.486	1.007
330.0	12.8580	640.8	1525.9	7186.3	0.489	74.5	7.70	5.513	1.095
340.0	14.6010	610.7	1594.5	8208.0	0.469	70.4	5.63	7.175	1.234
350.0	16.5290	574.7	1670.9	10,116.0	0.447	65.9	3.67	10.393	1.490
360.0	18.6660	527.6	1761.7	15,004.0	0.426	60.3	1.88	19.121	2.126
370.0	21.0440	451.4	1890.7	45,155.0	0.425	52.1	0.39	76.384	5.532

Table C.4 Thermophysical properties of water (vapor) on saturation line (based on temperature increments)

t	P_s	ρ_v	h_v	h_{fg}	c_{pv}	$k_v \times 10^2$	$\mu_v \times 10^6$	$\beta_{th} \times 10^3$	Pr_v
°C	MPa	kg/m³	kJ/kg	kJ/kg	J/kg K	W/m K	Pa s	1/K	—
5.0	0.0009	0.007	2510.1	2489.0	1889.4	1.734	9.336	3.619	1.017
10.0	0.0012	0.009	2519.2	2477.2	1894.7	1.762	9.461	3.559	1.017
15.0	0.0017	0.013	2528.3	2465.4	1900.2	1.792	9.592	3.502	1.017
20.0	0.0023	0.017	2537.4	2453.5	1905.9	1.823	9.727	3.447	1.017
25.0	0.0032	0.023	2546.5	2441.7	1911.8	1.855	9.867	3.395	1.017
30.0	0.0042	0.030	2555.5	2429.8	1918.0	1.889	10.010	3.345	1.017
35.0	0.0056	0.040	2564.5	2417.9	1924.5	1.924	10.157	3.298	1.016
40.0	0.0074	0.051	2573.5	2406.0	1931.4	1.960	10.308	3.253	1.016
45.0	0.0096	0.066	2582.4	2394.0	1938.8	1.998	10.461	3.210	1.015
50.0	0.0124	0.083	2591.3	2381.9	1946.8	2.037	10.616	3.170	1.015
55.0	0.0158	0.105	2600.1	2369.8	1955.4	2.077	10.774	3.132	1.014
60.0	0.0199	0.130	2608.8	2357.7	1964.8	2.119	10.935	3.096	1.014
65.0	0.0250	0.161	2617.5	2345.4	1975.0	2.162	11.097	3.063	1.014
70.0	0.0312	0.198	2626.1	2333.0	1986.2	2.207	11.260	3.033	1.013
75.0	0.0386	0.242	2634.6	2320.6	1998.5	2.253	11.426	3.004	1.013
80.0	0.0474	0.294	2643.0	2308.0	2012.0	2.301	11.592	2.979	1.014
85.0	0.0579	0.354	2651.3	2295.3	2026.7	2.351	11.760	2.956	1.014
90.0	0.0702	0.424	2659.5	2282.5	2042.9	2.402	11.929	2.935	1.015
95.0	0.0846	0.505	2667.6	2269.5	2060.7	2.455	12.099	2.917	1.016
100.0	0.1014	0.598	2675.6	2256.4	2080.0	2.510	12.269	2.902	1.017
105.0	0.1209	0.705	2683.4	2243.1	2101.2	2.566	12.440	2.890	1.019
110.0	0.1434	0.827	2691.1	2229.6	2124.4	2.625	12.612	2.881	1.021
115.0	0.1692	0.965	2698.6	2216.0	2149.6	2.685	12.784	2.875	1.024
120.0	0.1987	1.122	2705.9	2202.1	2177.0	2.747	12.956	2.872	1.027
130.0	0.2703	1.497	2720.1	2173.7	2238.9	2.877	13.301	2.876	1.035
140.0	0.3615	1.967	2733.4	2144.3	2310.9	3.014	13.647	2.894	1.046
150.0	0.4762	2.548	2745.9	2113.7	2393.9	3.160	13.992	2.926	1.060
160.0	0.6182	3.260	2757.4	2082.0	2488.3	3.313	14.337	2.973	1.077
170.0	0.7922	4.122	2767.9	2048.8	2594.4	3.475	14.681	3.036	1.096
180.0	1.0028	5.159	2777.2	2014.2	2712.9	3.645	15.025	3.116	1.118
190.0	1.2552	6.395	2785.3	1977.9	2844.3	3.824	15.370	3.215	1.143
200.0	1.5549	7.861	2792.0	1939.7	2989.5	4.011	15.715	3.334	1.171
210.0	1.9077	9.589	2797.3	1899.6	3150.3	4.209	16.061	3.476	1.202
220.0	2.3196	11.615	2800.9	1857.4	3328.9	4.417	16.411	3.645	1.237
230.0	2.7971	13.985	2802.9	1812.7	3528.5	4.638	16.765	3.845	1.276
240.0	3.3469	16.749	2803.0	1765.4	3753.7	4.873	17.125	4.084	1.319
250.0	3.9762	19.967	2800.9	1715.2	4010.5	5.126	17.495	4.370	1.369
260.0	4.6923	23.712	2796.6	1661.6	4307.5	5.403	17.877	4.715	1.425
270.0	5.5030	28.073	2789.7	1604.4	4656.3	5.711	18.277	5.136	1.490
280.0	6.4166	33.165	2779.9	1543.0	5073.1	6.061	18.700	5.658	1.565
290.0	7.4418	39.132	2766.7	1476.7	5582.1	6.471	19.154	6.316	1.652
300.0	8.5879	46.168	2749.6	1404.6	6219.7	6.965	19.651	7.166	1.755
310.0	9.8651	54.541	2727.9	1325.7	7044.9	7.584	20.207	8.299	1.877

Table C.4 Thermophysical properties of water (vapor) on saturation line (based on temperature increments)—cont'd

t	P_s	ρ_v	h_v	h_{fg}	c_{pv}	$k_v \times 10^2$	$\mu_v \times 10^6$	$\beta_{th} \times 10^3$	Pr_v
°C	MPa	kg/m³	kJ/kg	kJ/kg	J/kg K	W/m K	Pa s	1/K	—
320.0	11.2840	64.638	2700.6	1238.4	8158.9	8.391	20.846	9.871	2.027
330.0	12.8580	77.050	2666.0	1140.2	9752.6	9.494	21.606	12.181	2.219
340.0	14.6010	92.759	2621.8	1027.3	12,236.0	11.091	22.554	15.877	2.488
350.0	16.5290	113.610	2563.6	892.8	16,692.0	13.595	23.820	22.688	2.925
360.0	18.6660	143.900	2481.5	719.8	27,356.0	18.151	25.724	39.436	3.877
370.0	21.0440	201.840	2334.5	443.8	96,598.0	32.384	29.678	151.640	8.853

Table C.5 Specific volumes, specific enthalpies, and latent heat of vaporization of water (liquid and vapor) on saturation line (based on pressure increments)

p	t_s	v	v_v	h	h_v	h_{fg}
MPa	°C	m³/kg	m³/kg	kJ/kg	kJ/kg	kJ/kg
0.001	7.0	0.0010001	129.2080	29.33	2513.8	2484.5
0.002	17.5	0.0010012	67.006000	73.45	2533.2	2459.8
0.003	24.1	0.0010027	45.668000	101.00	2545.2	2444.2
0.004	29.0	0.0010040	34.803000	121.41	2554.1	2432.7
0.005	32.9	0.0010052	28.196000	137.77	2561.2	2423.4
0.006	36.2	0.0010064	23.742000	151.50	2567.1	2415.6
0.007	39.0	0.0010074	20.532000	163.38	2572.2	2408.8
0.008	41.5	0.0010089	18.106000	173.87	2576.7	2402.8
0.009	43.8	0.0010094	16.206000	183.28	2580.8	2397.5
0.010	45.8	0.0010102	14.676000	191.84	2584.4	2392.6
0.012	49.5	0.0010119	12.364000	206.94	2590.9	2384.0
0.014	52.6	0.0010133	10.696000	220.03	2596.4	2376.4
0.016	55.3	0.0010147	9.434800	231.60	2601.3	2369.7
0.018	57.8	0.0010160	8.447000	242.00	2605.7	2363.7
0.020	60.1	0.0010172	7.651500	251.46	2609.6	2358.1
0.022	62.2	0.0010183	6.996700	260.14	2613.2	2353.1
0.024	64.1	0.0010194	6.448300	268.18	2616.6	2348.4
0.026	65.9	0.0010204	5.981900	275.68	2619.7	2344.0
0.028	67.6	0.0010214	5.580400	282.70	2622.6	2339.9
0.030	69.1	0.0010223	5.230800	289.31	2625.3	2336.0
0.040	75.9	0.0010265	3.994900	317.65	2636.8	2319.2
0.050	81.3	0.0010301	3.241500	340.57	2646.0	2305.4
0.060	86.0	0.0010333	2.732900	359.93	2653.6	2293.7
0.070	90.0	0.0010361	2.365800	376.77	2660.2	2283.4

Continued

Table C.5 Specific volumes, specific enthalpies, and latent heat of vaporization of water (liquid and vapor) on saturation line (based on pressure increments)—cont'd

p	t_s	v	v_v	h	h_v	h_{fg}
MPa	°C	m³/kg	m³/kg	kJ/kg	kJ/kg	kJ/kg
0.080	93.5	0.0010387	2.087900	391.72	2666.0	2274.3
0.090	96.7	0.0010412	1.870100	405.21	2671.1	2265.9
0.100	99.6	0.0010434	1.694600	417.51	2675.7	2258.2
0.150	111.4	0.0010530	1.159700	467.13	2693.9	2226.8
0.200	120.2	0.0010608	0.885920	504.70	2706.9	2202.2
0.250	127.4	0.0010675	0.718810	535.40	2717.2	2181.8
0.300	133.5	0.0010735	0.605860	561.40	2725.5	2164.1
0.350	138.9	0.0010789	0.524250	584.30	2732.5	2148.2
0.400	143.6	0.0010839	0.462420	604.70	2738.5	2133.8
0.450	147.9	0.0010885	0.413920	623.20	2743.8	2120.6
0.500	151.9	0.0010928	0.374810	640.10	2748.5	2108.4
0.600	158.8	0.0011009	0.315560	670.40	2756.4	2086.0
0.700	1645.0	0.0011082	0.272740	697.10	2762.9	2065.8
0.800	170.4	0.0011150	0.240300	720.90	2768.4	2047.5
0.900	175.43	0.0011213	0.214840	742.60	2773.0	2030.4
1.000	179.9	0.0011274	0.194300	762.60	2777.0	2014.4
1.100	184.1	0.0011331	0.177390	781.10	2780.4	1999.3
1.200	188.0	0.0011386	0.163200	798.40	2783.4	1985.0
1.300	191.6	0.0011438	0.151120	814.70	2786.0	1971.3
1.400	195.0	0.0011489	0.140720	830.10	2788.4	1958.3
1.500	198.3	0.0011538	0.131650	844.70	2790.4	1945.7
1.600	201.3	0.0011586	0.123680	858.60	2792.2	1933.6
1.700	204.3	0.0011633	0.116610	871.80	2793.8	1922.0
1.800	207.1	0.0011678	0.110310	884.60	2795.1	1910.5
1.900	209.8	0.0011722	0.104640	896.80	2796.4	1899.6
2.000	212.8	0.0011766	0.099530	908.60	2797.4	1888.8
2.100	214.8	0.0011808	0.094880	919.90	2798.3	1878.4
2.200	217.2	0.0011850	0.090640	930.90	2799.1	1868.2
2.300	219.5	0.0011891	0.086760	941.60	2799.8	1858.2
2.400	221.7	0.0011932	0.083190	951.90	2800.4	1848.5
2.500	223.9	0.0011972	0.079900	962.00	2800.8	1838.8
2.600	226.0	0.0012011	0.076850	971.70	2801.2	1829.5
2.700	228.1	0.0012050	0.074020	981.20	2801.5	1820.3
2.800	230.0	0.0012088	0.071380	990.50	2801.7	1811.2
2.900	232.0	0.0012126	0.068920	999.50	2801.8	1802.3
3.000	233.9	0.0012163	0.066620	1008.40	2801.9	1793.5
3.100	235.7	0.0012200	0.064460	1017.00	2801.9	1784.9
3.200	237.4	0.0012237	0.062430	1025.50	2801.8	1776.3
3.300	239.2	0.0012273	0.060520	1033.70	2801.7	1768.0

Table C.5 Specific volumes, specific enthalpies, and latent heat of vaporization of water (liquid and vapor) on saturation line (based on pressure increments)—cont'd

p	t_s	v	v_v	h	h_v	h_{fg}
MPa	°C	m³/kg	m³/kg	kJ/kg	kJ/kg	kJ/kg
3.400	240.9	0.0012310	0.058720	1041.80	2801.5	1759.7
3.500	242.5	0.0012345	0.057020	1049.80	2801.3	1751.5
3.600	244.2	0.0012381	0.055400	1057.60	2801.0	1743.4
3.700	245.8	0.0012416	0.053880	1065.30	2800.7	1735.4
3.800	247.3	0.0012451	0.052430	1072.80	2800.3	1727.5
3.900	248.8	0.0012486	0.051050	1080.20	2799.9	1719.7
4.000	250.3	0.0012521	0.049740	1087.50	2799.4	1711.9
4.100	251.8	0.0012555	0.048490	1094.60	2798.9	1704.3
4.200	253.2	0.0012589	0.047290	1101.70	2798.4	1696.7
4.300	254.7	0.0012623	0.046150	1108.60	2797.8	1689.2
4.400	256.1	0.0012657	0.045060	1115.50	2797.2	1681.7
4.500	257.4	0.0012691	0.044020	1122.20	2796.5	1674.3
4.600	258.8	0.0012725	0.043020	1128.90	2795.9	1667.0
4.700	260.1	0.0012758	0.042060	1135.40	2795.2	1659.8
4.800	261.4	0.0012792	0.041140	1141.90	2794.4	1652.5
4.900	262.7	0.0012825	0.040260	1148.30	2793.6	1645.3
5.000	263.9	0.0012858	0.039410	1154.60	2792.8	1638.2
5.200	266.4	0.0012925	0.037800	1167.00	2791.1	1624.1
5.400	268.8	0.0012990	0.036310	1179.10	2789.3	1610.2
5.600	271.1	0.0013056	0.034920	1191.00	2787.4	1596.4
5.800	273.4	0.0013122	0.033630	1202.60	2785.4	1582.8
6.000	275.6	0.0013187	0.032410	1213.90	2783.3	1569.4
6.200	277.7	0.0013252	0.031270	1225.10	2781.1	1556.0
6.400	279.8	0.0013318	0.030200	1236.00	2778.8	1542.8
6.600	281.9	0.0013383	0.029200	1246.80	2776.4	1529.6
6.800	283.9	0.0013448	0.028240	1257.30	2773.9	1516.6
7.000	285.8	0.0013514	0.027340	1267.70	2771.4	1503.7
7.200	287.7	0.0013579	0.026490	1278.00	2768.7	1490.7
7.400	289.6	0.0013645	0.025680	1288.00	2766.1	1478.1
7.600	291.4	0.0013711	0.024920	1298.00	2763.3	1465.3
7.800	293.2	0.0013777	0.024190	1307.80	2760.4	1452.6
8.000	295.0	0.0013843	0.023490	1317.50	2757.5	1440.0
8.200	296.7	0.0013909	0.022830	1327.00	2754.5	1427.5
8.400	298.4	0.0013976	0.022200	1336.50	2751.4	1414.9
8.600	300.1	0.0014043	0.021590	1345.80	2748.3	1402.5
8.800	301.7	0.0014111	0.021010	1355.10	2745.1	1390.0
9.000	303.3	0.0014179	0.020460	1364.20	2741.8	1377.6
9.200	305.0	0.0014247	0.019930	1373.20	2738.5	1365.3
9.400	306.5	0.0014316	0.019420	1382.20	2735.1	1352.9

Continued

Table C.5 Specific volumes, specific enthalpies, and latent heat of vaporization of water (liquid and vapor) on saturation line (based on pressure increments)—cont'd

p	t_s	v	v_v	h	h_v	h_{fg}
MPa	°C	m³/kg	m³/kg	kJ/kg	kJ/kg	kJ/kg
9.600	308.0	0.0014385	0.018930	1391.10	2731.6	1340.5
9.800	309.5	0.0014455	0.018450	1399.90	2728.0	1328.1
10.000	311.0	0.0014526	0.018000	1406.60	2724.4	1315.8
10.200	312.4	0.0014597	0.017560	1417.30	2720.8	1303.5
10.400	313.9	0.0014668	0.017140	1425.80	2717.1	1291.3
10.600	315.3	0.0014740	0.016740	1434.40	2713.2	1278.8
10.800	316.7	0.0014813	0.016350	1442.80	2709.4	1266.6
11.000	318.0	0.0014887	0.015970	1451.20	2705.4	1254.2
11.200	319.4	0.0014961	0.015600	1459.60	2701.5	1241.9
11.400	320.7	0.0015036	0.015250	1467.90	2697.3	1229.4
11.600	322.1	0.0015112	0.014900	1476.20	2693.2	1217.0
11.800	323.4	0.0015189	0.014570	1484.40	2689.0	1204.6
12.000	324.6	0.0015267	0.014250	1492.60	2684.8	1192.2
12.200	325.9	0.0015345	0.013940	1500.70	2680.4	1179.7
12.400	327.2	0.0015425	0.013630	1508.80	2676.0	1167.2
12.600	328.4	0.0015506	0.013340	1516.90	2671.6	1154.7
12.800	329.6	0.0015588	0.013050	1524.90	2667.0	1142.1
13.000	330.8	0.0015670	0.012770	1533.00	2662.4	1129.4
13.200	332.0	0.0015755	0.012500	1541.00	2657.7	1116.7
13.400	333.2	0.0015840	0.012240	1548.90	2653.0	1104.1
13.600	334.3	0.0015927	0.011990	1556.90	2648.2	1091.3
13.800	335.5	0.0016015	0.011740	1564.80	2643.3	1078.5
14.000	336.6	0.0016104	0.011490	1572.80	2638.3	1065.5
14.200	337.8	0.0016195	0.011260	1580.70	2633.2	1052.5
14.400	338.9	0.0016288	0.011020	1588.60	2628.1	1039.5
14.600	340.0	0.0016382	0.010800	1596.50	2622.9	1026.4
14.800	341.0	0.0016483	0.010570	1604.30	2617.1	1012.8
15.000	342.1	0.0016580	0.010350	1612.20	2611.6	999.4
15.200	343.2	0.0016680	0.010140	1620.00	2606.1	986.1
15.400	344.2	0.0016782	0.009930	1627.90	2600.4	972.5
15.600	345.3	0.0016886	0.009726	1635.70	2594.6	958.9
15.800	346.3	0.0016992	0.009526	1643.60	2588.7	945.1
16.000	347.3	0.0017101	0.009330	1651.50	2582.7	931.2
16.200	348.3	0.0017212	0.009138	1659.40	2576.6	917.2
16.400	349.3	0.0017327	0.008949	1667.40	2570.4	903.0
16.600	350.3	0.0017444	0.008763	1675.40	2564.0	888.6
16.800	351.3	0.0017565	0.008581	1683.50	2557.5	874.0
17.000	352.3	0.0017690	0.008401	1691.60	2550.8	859.2
17.200	353.2	0.0017818	0.008223	1699.80	2544.0	844.2

Table C.5 Specific volumes, specific enthalpies, and latent heat of vaporization of water (liquid and vapor) on saturation line (based on pressure increments)—cont'd

p	t_s	v	v_v	h	h_v	h_{fg}
MPa	°C	m³/kg	m³/kg	kJ/kg	kJ/kg	kJ/kg
17.400	354.2	0.0017951	0.006048	1708.10	2536.9	828.8
17.600	355.1	0.0018069	0.007875	1716.40	2529.7	813.3
17.800	356.0	0.0018231	0.007704	1724.90	2522.2	797.3
18.000	357.0	0.0018380	0.007534	1733.40	2514.4	781.0
18.200	357.9	0.0018534	0.007366	1742.10	2506.3	764.2
18.400	358.8	0.0018696	0.007198	1750.80	2497.9	747.1
18.600	359.7	0.0018865	0.007032	1759.80	2489.1	729.3
18.800	360.6	0.0019043	0.006866	1768.90	2479.8	710.9
19.000	361.4	0.0019231	0.006700	1778.20	2470.1	691.9
19.200	362.3	0.0019430	0.006534	1787.70	2459.8	672.1
19.400	363.2	0.0019642	0.006369	1797.50	2449.1	651.6
19.600	364.0	0.0019869	0.006204	1807.50	2437.8	630.3
19.800	364.9	0.0020110	0.006039	1817.90	2426.1	608.2
20.000	365.7	0.0020380	0.006873	1828.80	2413.8	585.0
20.200	366.5	0.0020660	0.005706	1840.10	2400.9	560.8
20.400	367.4	0.0020980	0.005537	1851.90	2387.3	535.4
20.600	368.2	0.0021340	0.005365	1864.50	2372.8	508.3
20.800	369.0	0.0021730	0.005189	1877.90	2357.2	479.3
21.000	369.8	0.0022180	0.005006	1892.20	2340.2	448.0
21.200	370.60	0.0022680	0.004780	1907.60	2321.3	413.7
21.400	371.40	0.0023270	0.004609	1924.70	2299.8	375.1
21.600	372.2	0.0024080	0.004382	1946.00	2274.2	328.2
21.800	372.9	0.0025100	0.004116	1971.00	2241.6	270.6
22.000	373.7	0.0026750	0.003757	2007.70	2192.5	184.8

Table C.6 Specific volumes and specific enthalpies of water at various temperatures and pressures

T	υ	h	υ	h	υ	h
°C	m³/kg	kJ/kg	m³/kg	kJ/kg	m³/kg	kJ/kg
	$p = 0.1$ MPa		$p = 0.2$ MPa		$p = 0.3$ MPa	
0	0.0010002	0.0	0.0010001	0.2	0.0010001	0.3
10	0.0010003	42.1	0.0010002	42.2	0.0010001	42.3
20	0.0010017	84.0	0.0010016	84.0	0.0010016	84.1
30	0.0010043	125.8	0.0010042	125.8	0.0010042	125.9
40	0.0010078	167.5	0.0010077	167.6	0.0010077	167.7

Continued

Table C.6 Specific volumes and specific enthalpies of water at various temperatures and pressures—cont'd

T	υ	h	υ	h	υ	h
°C	m³/kg	kJ/kg	m³/kg	kJ/kg	m³/kg	kJ/kg
50	0.0010121	209.3	0.0010120	209.4	0.0010120	209.5
60	0.0010171	251.2	0.0010171	251.2	0.0010170	251.3
70	0.0010228	293.0	0.0010228	293.1	0.0010227	293.2
80	0.0010292	335.0	0.0010291	335.0	0.0010291	335.1
90	0.0010361	377.0	0.0010361	377.0	0.0010360	377.1
100	—	—	0.0010437	419.1	0.0010436	419.2
110	—	—	0.0010518	461.4	0.0010518	461.4
120	—	—	0.0010606	503.7	0.0010606	503.8
130	—	—			0.0010700	546.3
	$p = 0.4$ MPa		$p = 0.6$ MPa		$p = 0.8$ MPa	
0	0.0010000	0.4	0.0009999	0.6	0.0009998	0.8
10	0.0010001	42.4	0.0010000	42.6	0.0009999	42.8
20	0.0010015	84.2	0.0010014	84.4	0.0010014	84.6
30	0.0010041	126.0	0.0010040	126.2	0.0010040	126.4
40	0.0010076	167.8	0.0010075	168.0	0.0010075	168.2
50	0.0010119	209.6	0.0010118	209.8	0.0010118	209.9
60	0.0010170	251.4	0.0010169	251.6	0.0010168	251.7
70	0.0010227	293.3	0.0010226	293.4	0.0010225	293.6
80	0.0010290	335.2	0.0010289	335.4	0.0010288	335.5
90	0.0010360	377.2	0.0010359	377.3	0.0010358	377.5
100	0.0010436	419.3	0.0010434	419.4	0.0010433	419.6
110	0.0010517	461.5	0.0010516	461.6	0.0010515	461.8
120	0.0010605	503.9	0.0010604	504.0	0.0010603	504.1
130	0.0010699	546.4	0.0010698	546.5	0.0010697	546.7
140	0.0010800	589.1	0.0010799	589.3	0.0010798	589.4
150	—	—	0.0010907	632.2	0.0010906	632.4
160	—	—	—	—	0.0011021	675.6
170	—	—	—	—	0.0011144	719.1
	$p = 1.0$ MPa		$p = 1.5$ MPa		$p = 2.0$ MPa	
0	0.0009997	1.0	0.0009995	1.5	0.0009992	2.0
10	0.0009998	43.0	0.0009995	43.5	0.0009993	43.9
20	0.0010013	84.8	0.0010010	85.3	0.0010008	85.7
30	0.0010039	126.6	0.0010036	127.0	0.0010034	127.5
40	0.0010074	168.3	0.0010071	168.8	0.0010069	169.2
50	0.0010117	210.1	0.0010114	210.6	0.0010112	211.0
60	0.0010167	251.9	0.0010165	252.3	0.0010162	252.7
70	0.0010224	293.8	0.0010222	294.2	0.0010219	294.6
80	0.0010287	335.7	0.0010285	336.1	0.0010282	336.5
90	0.0010357	377.7	0.0010354	378.0	0.0010352	378.4

Table C.6 Specific volumes and specific enthalpies of water at various temperatures and pressures—cont'd

T	v	h	v	h	v	h
°C	m³/kg	kJ/kg	m³/kg	kJ/kg	m³/kg	kJ/kg
100	0.0010432	419.7	0.0010430	420.1	0.0010427	420.5
110	0.0010514	461.9	0.0010511	462.3	0.0010508	462.7
120	0.0010602	504.3	0.0010599	504.6	0.0010596	505.0
130	0.0010696	546.8	0.0010693	547.1	0.0010690	547.5
140	0.0010796	589.5	0.0010793	589.8	0.0010790	590.2
150	0.0010904	632.5	0.0010901	632.8	0.0010897	633.1
160	0.0011019	675.7	0.0011016	676.0	10.0011012	676.3
170	0.0011143	719.21	0.0011139	719.5	0.0011135	719.8
180	—	—	0.0011271	763.4	0.0011266	763.6
190	—	—	0.0011413	807.6	0.0011408	807.9
200	—	—	—	—	0.0011560	852.6
210	—	—	—	—	0.0011725	897.8
	p = 3.0 MPa		*p* = 4.0 MPa		*p* = 5.0 MPa	
0	0.0009987	3.0	0.0009982	4.0	0.0009977	5.1
10	0.0009988	44.9	0.0009984	45.9	0.0009979	46.9
20	0.0010004	86.7	0.0009999	87.6	0.0009995	88.6
30	0.0010030	128.4	0.0010025	129.3	0.0010021	130.2
40	0.0010065	170.1	0.0010060	171.0	0.0010056	171.9
50	0.0010108	211.8	0.0010103	212.7	0.0010099	213.6
60	0.0010158	253.6	0.0010153	254.4	0.0010149	255.3
70	0.0010215	295.4	0.0010210	296.2	0.0010205	297.0
80	0.0010278	337.3	0.0010273	338.1	0.0010268	338.8
90	0.0010347	379.2	0.0010342	380.0	0.0010337	380.7
100	0.0010422	421.2	0.0010417	422.0	0.0010412	422.7
110	0.0010503	463.4	0.0010498	464.1	0.0010492	464.8
120	0.0010590	505.7	0.0010584	506.4	0.0010579	507.1
130	0.0010684	548.2	0.0010677	548.8	0.0010671	549.5
140	0.0010783	590.8	0.0010777	591.5	0.0010771	592.1
150	0.0010890	633.7	0.0010883	634.3	0.0010877	635.0
160	0.0011005	676.9	0.0010997	677.5	0.0010990	678.0
170	0.0011127	720.3	0.0011119	720.9	0.0011111	721.4
180	0.0011258	764.1	0.0011249	764.6	0.0011241	765.2
190	0.0011399	808.3	0.0011389	808.8	0.0011380	809.3
200	0.0011550	853.0	0.0011540	853.4	0.0011530	853.8
210	0.0011714	898.1	0.0011702	898.5	0.0011691	898.8
220	0.0011891	943.9	0.0011878	944.2	0.0011866	944.4
230	0.0012084	990.3	0.0012070	990.5	0.0012056	990.7
240	—	—	0.0012280	1037.7	0.0012264	1037.8
250	—	—	0.0012512	1085.8	0.0012494	1085.8
260	—	—	—	—	0.0012750	1135.0

Continued

Table C.6 Specific volumes and specific enthalpies of water at various temperatures and pressures—cont'd

T	v	h	v	h	v	h
°C	m³/kg	kJ/kg	m³/kg	kJ/kg	m³/kg	kJ/kg
	$p = 6.0$ MPa		$p = 7.0$ MPa		$p = 8.0$ MPa	
0	0.0009972	6.1	0.0009967	7.1	0.0009962	8.1
10	0.0009974	47.8	0.0009970	48.8	0.0009965	49.8
20	0.0009990	89.5	0.0009986	90.4	0.0009981	91.4
30	0.0010016	131.1	0.0010012	132.0	0.0010008	132.9
40	0.0010051	172.7	0.0010047	173.6	0.0010043	174.5
50	0.0010094	214.4	0.0010090	215.3	0.0010086	216.1
60	0.0010144	256.1	0.0010140	256.9	0.0010135	257.8
70	0.0010201	297.8	0.0010196	298.7	0.0010192	299.5
80	0.0010263	339.6	0.0010259	340.4	0.0010254	341.2
90	0.0010332	381.5	0.0010327	382.3	0.0010322	383.1
100	0.0010406	423.5	0.0010401	424.2	0.0010396	425.0
110	0.0010487	465.6	0.0010481	466.3	0.0010476	467.0
120	0.0010573	507.8	0.0010567	508.5	0.0010562	509.2
130	0.0010665	550.2	0.0010660	550.9	0.0010654	551.6
140	0.0010764	592.8	0.0010758	593.4	0.0010752	594.1
150	0.0010870	635.6	0.0010863	636.2	0.0010856	636.8
160	0.0010983	678.6	0.0010976	679.2	10.0010968	679.8
170	0.0011103	722.0	0.0011096	722.6	0.0011088	723.1
180	0.0011232	765.7	0.0011224	766.2	0.0011216	766.7
190	0.0011371	809.7	0.0011362	810.2	0.0011353	810.7
200	0.0011519	854.2	0.0011510	854.6	0.0011500	855.1
210	0.0011680	899.2	0.0011669	899.6	0.0011658	899.9
220	0.0011853	944.7	0.0011841	945.0	0.0011829	945.3
230	0.0012042	990.9	0.0012028	991.1	0.0012015	991.4
240	0.0012249	1037.9	0.0012233	1038	0.0012218	1038.2
250	0.0012476	1085.8	0.0012458	1085.8	0.001244	1085.8
260	0.0012729	1134.8	0.0012708	1134.7	0.0012687	1134.6
270	0.0013013	1185.2	0.0012988	1184.9	0.0012964	1184.6
280	—	—	0.0013307	1236.7	0.0013277	1236.2
290	—	—	—	—	0.0013639	1289.8
	$p = 9.0$ MPa		$p = 10.0$ MPa		$p = 11.0$ MPa	
0	0.0009958	9.1	0.0009953	10.1	0.0009948	11.1
10	0.0009960	50.7	0.0009956	51.7	0.0009951	52.7
20	0.0009977	92.3	0.0009972	93.2	0.0009968	94.2
30	0.0010003	133.8	0.0009999	134.7	0.0009995	135.7
40	0.0010038	175.4	0.0010034	176.3	0.0010030	177.2

Table C.6 Specific volumes and specific enthalpies of water at various temperatures and pressures—cont'd

T	v	h	v	h	v	h
°C	m³/kg	kJ/kg	m³/kg	kJ/kg	m³/kg	kJ/kg
50	0.0010081	217.0	0.0010077	217.8	0.0010073	218.7
60	0.0010131	258.6	0.0010126	259.4	0.0010122	260.3
70	0.0010187	300.3	0.0010182	301.1	0.0010178	301.9
80	0.0010249	342.0	0.0010244	342.8	0.0010240	343.6
90	0.0010317	383.8	0.0010312	384.6	0.0010308	385.4
100	0.0010391	425.8	0.0010386	426.5	0.0010381	427.2
110	0.0010471	467.8	0.0010465	468.5	0.0010460	469.2
120	0.0010556	509.9	0.0010551	510.6	0.0010545	511.3
130	0.0010648	552.2	0.0010642	552.9	0.0010636	553.6
140	0.0010745	594.7	0.0010739	595.4	0.0010733	596.1
150	0.0010850	637.5	0.0010843	638.1	0.0010837	638.7
160	0.0010961	680.4	0.0010954	681.0	0.0010947	681.6
170	0.0011080	723.7	0.0011072	724.2	0.0011065	724.8
180	0.0011207	767.2	0.0011199	767.8	0.0011191	768.3
190	0.0011344	811.2	0.0011335	811.6	0.0011326	812.1
200	0.0011490	855.5	0.0011480	855.9	0.0011470	856.4
210	0.0011647	900.3	0.0011636	900.7	0.0011626	901.1
220	0.0011817	945.6	0.0011805	946.0	0.0011793	946.3
230	0.0012001	991.6	0.0011988	991.8	0.0011975	992.1
240	0.0012202	1038.3	0.0012188	1038.4	0.0012173	1038.6
250	0.0012423	1085.9	0.0012406	1085.9	0.0012389	1085.9
260	0.0012667	1134.4	0.0012648	1134.3	0.0012628	1134.3
270	0.0012940	1184.3	0.0012917	1184.0	0.0012894	1183.8
280	0.0013249	1235.6	0.0013221	1235.2	0.0013194	1234.7
290	0.0013604	1289.0	0.0013570	1288.2	0.0013536	1287.5
300	0.0014022	1344.9	0.0013978	1343.7	0.0013936	1342.6
310	—	—	0.0014472	1402.6	0.0014416	1400.9
	$p = 12.0$ MPa		$p = 13.0$ MPa		$p = 14.0$ MPa	
0	0.0009943	12.1	0.0009938	13.1	0.0009933	14.1
10	0.0009947	53.6	0.0009942	54.6	0.0009938	55.6
20	0.0009964	95.1	0.0009959	96.0	0.0009955	97
30	0.0009991	136.6	0.0009986	137.4	0.0009982	138.4
40	0.0010026	178.1	0.0010021	178.9	0.0010017	179.8
50	0.0010068	219.6	0.0010064	220.4	0.001006	221.3
60	0.0010118	261.1	0.0010113	262.0	0.0010109	262.8
70	0.0010174	302.7	0.0010169	303.6	0.0010164	304.4
80	0.0010235	344.4	0.0010231	345.2	0.0010226	346.0
90	0.0010303	386.2	0.0010298	386.9	0.0010293	387.7

Continued

Table C.6 Specific volumes and specific enthalpies of water at various temperatures and pressures—cont'd

T	υ	h	υ	h	υ	h
°C	m³/kg	kJ/kg	m³/kg	kJ/kg	m³/kg	kJ/kg
100	0.0010376	428.0	0.0010371	428.8	0.0010366	429.5
110	0.0010455	470.0	0.0010450	470.7	0.0010444	471.4
120	0.001054	512.0	0.0010534	512.8	0.0010529	513.5
130	0.001063	554.3	0.0010624	555.0	0.0010619	555.7
140	0.0010727	596.7	0.0010721	597.4	0.0010715	598.0
150	0.001083	639.3	0.0010824	640.0	0.0010817	640.6
160	0.001094	682.2	0.0010933	682.8	0.0010926	683.4
170	0.0011058	725.4	0.0011050	725.9	0.0011043	726.5
180	0.0011183	768.8	0.0011175	769.4	0.0011167	769.9
190	0.0011317	812.6	0.0011308	813.1	0.00113	813.6
200	0.0011461	856.8	0.0011451	857.2	0.0011442	857.7
210	0.0011615	901.4	0.0011605	901.8	0.0011594	902.2
220	0.0011782	946.6	0.0011770	946.9	0.0011759	947.2
230	0.0011962	992.3	0.0011949	992.6	0.0011936	992.8
240	0.0012158	1038.8	0.0012144	1038.9	0.0012129	1039.1
250	0.0012373	1086.0	0.0012356	1086.1	0.0012340	1086.1
260	0.0012609	1134.2	0.0012590	1134.1	0.0012572	1134.1
270	0.0012872	1183.5	0.0012850	1183.3	0.0012828	1183.1
280	0.0013167	1234.3	0.0013141	1233.9	0.0013115	1233.5
290	0.0013504	1286.8	0.0013472	1286.1	0.0013441	1285.5
300	0.0013895	1341.5	0.0013855	1340.5	0.0013816	1339.5
310	0.0014362	1399.3	0.0014310	1397.8	0.0014260	1396.4
320	0.0014941	1461.5	0.0014869	1459.2	0.0014801	1457.0
330	—	—	0.0015600	1526.9	0.0015497	1523.5
	$p = 15.0$ MPa		$p = 16.0$ MPa		$p = 17.0$ MPa	
0	0.0009928	15.1	0.0009924	16.1	0.0009919	17.1
10	0.0009933	56.5	0.0009928	57.5	0.0009924	58.4
20	0.0009950	97.9	0.0009946	98.8	0.0009942	99.7
30	0.0009978	139.3	0.0009973	140.2	0.0009969	141.1
40	0.0010013	180.7	0.0010008	181.6	0.0010004	182.4
50	0.0010055	222.1	0.0010051	223.0	0.0010047	223.8
60	0.0010105	263.6	0.0010100	264.5	0.0010096	265.3
70	0.0010160	305.2	0.0010156	306.0	0.0010151	306.8
80	0.0010221	346.8	0.0010217	347.6	0.0010212	348.4
90	0.0010289	388.5	0.0010284	389.3	0.0010279	390.0
100	0.0010361	430.3	0.0010356	431.0	0.0010351	431.8
110	0.0010439	472.2	0.0010434	472.9	0.0010429	473.6
120	0.0010523	514.2	0.0010518	514.9	0.0010512	515.6
130	0.0010613	556.4	0.0010607	557.0	0.0010602	557.7

Table C.6 Specific volumes and specific enthalpies of water at various temperatures and pressures—cont'd

T	v	h	v	h	v	h
°C	m³/kg	kJ/kg	m³/kg	kJ/kg	m³/kg	kJ/kg
140	0.0010709	598.7	0.0010703	599.4	0.0010697	600.0
150	0.0010811	641.3	0.0010804	641.9	0.0010798	642.5
160	0.0010919	684.0	0.0010912	684.6	0.0010906	685.2
170	0.0011035	727.1	0.0011028	727.7	0.0011021	728.2
180	0.0011159	770.4	0.0011151	771.0	0.0011143	771.5
190	0.0011291	814.1	0.0011282	814.6	0.0011274	815.1
200	0.0011432	858.1	0.0011423	858.6	0.0011414	859.0
210	0.0011584	902.6	0.0011574	903.0	0.0011564	903.4
220	0.0011748	947.6	0.0011736	947.9	0.0011725	948.3
230	0.0011924	993.1	0.0011912	993.4	0.0011899	993.7
240	0.0012115	1039.3	0.0012101	1039.5	0.0012088	1039.7
250	0.0012324	1086.2	0.0012308	1086.3	0.0012293	1086.4
260	0.0012553	1134.0	0.0012535	1134.0	0.0012517	1134.0
270	0.0012807	1182.9	0.0012786	1182.8	0.0012765	1182.6
280	0.0013090	1233.1	0.0013065	1232.8	0.0013041	1232.4
290	0.0013410	1284.9	0.0013381	1284.3	0.0013352	1283.8
300	0.0013779	1338.6	0.0013742	1337.7	0.0013707	1336.9
310	0.0014212	1395.0	0.0014165	1393.7	0.0014120	1392.4
320	0.0014736	1455.0	0.0014674	1453.0	0.0014615	1451.2
330	0.0015402	1520.3	0.0015312	1517.3	0.0015229	1514.6
340	0.0016323	1594.6	0.0016175	1589.6	0.0016042	1585.0
350	—	—	—	—	0.0017286	1668.7
	$p = 18.0$ MPa		$p = 19.0$ MPa		$p = 20.0$ MPa	
0	0.0009914	18.1	0.0009909	19.1	0.0009904	20.1
10	0.0009919	59.4	0.0009915	60.4	0.0009910	61.3
20	0.0009937	100.7	0.0009933	101.6	0.0009929	102.5
30	0.0009965	142.0	0.0009960	142.9	0.0009956	143.8
40	0.0010000	183.3	0.0009996	184.2	0.0009992	185.1
50	0.0010043	224.7	0.0010038	225.6	0.0010034	226.4
60	0.0010092	266.1	0.0010087	267.0	0.0010083	267.8
70	0.0010147	307.6	0.0010142	308.4	0.0010138	309.3
80	0.0010208	349.2	0.0010203	350.0	0.0010199	350.8
90	0.0010274	390.8	0.0010270	391.6	0.0010265	392.4
100	0.0010346	432.5	0.0010342	433.3	0.0010337	434.0
110	0.0010424	474.4	0.0010419	475.1	0.0010414	475.8
120	0.0010507	516.3	0.0010502	517.0	0.0010496	517.7
130	0.0010596	558.4	0.001059	559.1	0.0010585	559.8
140	0.0010691	600.7	0.0010685	601.4	0.0010679	602.0

Continued

Table C.6 Specific volumes and specific enthalpies of water at various temperatures and pressures—cont'd

T	v	h	v	h	v	h
°C	m³/kg	kJ/kg	m³/kg	kJ/kg	m³/kg	kJ/kg
150	0.0010792	643.2	0.0010785	643.8	0.0010779	644.4
160	0.0010899	685.9	0.0010892	686.5	0.0010886	687.1
170	0.0011014	728.8	0.0011006	729.4	0.0010999	730.0
180	0.0011136	772.0	0.0011128	772.6	0.0011120	773.1
190	0.0011266	815.6	0.0011257	816.1	0.0011249	816.6
200	0.0011405	859.5	0.0011396	860.0	0.0011387	860.4
210	0.0011554	903.8	0.0011545	904.2	0.0011534	904.7
220	0.0011714	948.6	0.0011703	949.0	0.0011693	949.3
230	0.0011887	993.9	0.0011875	994.2	0.0011863	994.5
240	0.0012074	1039.9	0.0012061	1040.1	0.0012047	1040.3
250	0.0012277	1086.5	0.0012262	1086.7	0.0012247	1086.8
260	0.0012500	1134.0	0.0012483	1134.0	0.0012466	1134.1
270	0.0012745	1182.5	0.0012725	1182.4	0.0012706	1182.3
280	0.0013017	1232.1	0.0012994	1231.9	0.0012971	1231.6
290	0.0013324	1283.2	0.0013296	1282.8	0.0013269	1282.3
300	0.0013672	1336.1	0.0013639	1335.3	0.0013606	1334.6
310	0.0014077	1391.3	0.0014035	1390.1	0.0013994	1389.1
320	0.0014558	1449.5	0.0014503	1447.9	0.0014450	1446.3
330	0.0015150	1512.0	0.0015075	1509.6	0.0015003	1507.3
340	0.0015920	1580.9	0.0015807	1577.2	0.0015703	1573.7
350	0.0017040	1660.9	0.001683	1654.2	0.0016660	1648.4
355	0.001790	1710.2	0.001757	1699.8	0.001731	1691.4
360	—	—	0.001871	1756.8	0.001823	1742.0
365	—	—	—	—	0.001991	1813.2

Table C.7 Specific volumes and specific enthalpies of steam at various temperatures and pressures

T	v	h	v	h	v	h
°C	m³/kg	kJ/kg	m³/kg	kJ/kg	m³/kg	kJ/kg
	$p = 0.1$ MPa		$p = 0.15$ MPa		$p = 0.2$ MPa	
100	1.69600	2676.5	—	—	—	—
110	1.74500	2696.7	—	—	—	—
120	1.79300	2716.8	1.18800	2711.7	—	—
130	1.84100	2736.8	1.22100	2732.2	0.91040	2727.5
140	1.88900	2756.6	1.25300	2752.5	0.93520	2748.4
150	1.93700	2776.4	1.28600	2772.7	0.95980	2769.0
160	1.98400	2796.2	1.31800	2792.9	0.98420	2789.5
170	2.03100	2816.0	1.34900	2812.9	1.00800	2809.8
180	2.07800	2835.7	1.38100	2832.9	1.03200	2830.1

Table C.7 Specific volumes and specific enthalpies of steam at various temperatures and pressures—cont'd

T	v	h	v	h	v	h
°C	m³/kg	kJ/kg	m³/kg	kJ/kg	m³/kg	kJ/kg
190	2.12500	2855.4	1.41300	2852.8	1.05600	2850.3
200	2.17200	2875.2	1.44400	2872.8	1.08000	2870.4
210	2.21900	2894.9	1.47600	2892.7	1.10400	2890.5
220	2.26600	2914.7	1.50700	2912.6	1.12800	2910.6
230	2.31300	2934.5	1.53900	2932.6	1.15200	2930.7
240	2.35900	2954.3	1.57000	2952.6	1.17500	2950.8
250	2.40600	2974.2	1.60100	2972.5	1.19900	2970.8
260	2.45300	2994.1	1.63200	2992.5	1.22200	2990.9
270	2.49900	3014.0	1.66400	3012.5	1.24600	3011.1
280	2.54600	3034.0	1.69500	3032.6	1.26900	3031.2
290	2.59200	3054.0	1.72600	3052.7	1.29300	3051.4
300	2.63900	3074.1	1.75700	3072.9	1.31600	3071.6
310	2.68500	3094.3	1.78800	3093.1	1.34000	3091.9
320	2.73200	3114.4	1.81900	3113.3	1.36300	3112.2
330	2.77800	3134.7	1.85000	3133.6	1.38600	3132.5
340	2.82400	3155.1	1.88100	3153.9	1.41000	3152.9
350	2.87100	3175.3	1.91200	3174.3	1.43300	3173.3
360	2.91700	3195.7	1.94300	3194.8	1.45600	3193.8
370	2.96400	3216.2	1.97400	3215.3	1.47900	3214.4
380	3.01000	3236.7	2.00500	3235.9	1.50300	3235.0
390	3.05600	3257.3	2.03600	3256.5	1.52600	3255.7
400	3.10300	3278.0	2.06700	3277.2	1.54900	3276.4
410	3.14900	3298.7	2.09800	3297.9	1.57200	3297.2
420	3.19500	3319.5	2.12900	3318.7	1.59600	3318.0
430	3.24200	3340.3	2.16000	3339.6	1.61900	3338.9
440	3.28800	3361.2	2.19100	3360.5	1.64200	3359.8
450	3.33400	3382.2	2.22200	3381.5	1.66500	3380.8
	$p = 0.25$ MPa		$p = 0.3$ MPa		$p = 0.4$ MPa	
130	0.72410	2722.7	—	—	—	—
140	0.74430	2744.0	0.61700	2739.6	—	—
150	0.76440	2765.1	0.63400	2761.2	0.47080	2752.9
160	0.78420	2786.0	0.65080	2782.4	0.48390	2775.0
170	0.80340	2806.7	0.66740	2803.4	0.49670	2796.8
180	0.82330	2827.2	0.68380	2824.3	0.50940	2818.3
190	0.84270	2847.6	0.70020	2845.0	0.52190	2839.5
200	0.86200	2868.0	0.71640	2865.6	0.53430	2860.6
210	0.88120	2888.3	0.73250	2886.1	0.54660	2881.5
220	0.90040	2908.5	0.74860	2906.5	0.55880	2902.2

Continued

Table C.7 Specific volumes and specific enthalpies of steam at various temperatures and pressures—cont'd

T	v	h	v	h	v	h
°C	m³/kg	kJ/kg	m³/kg	kJ/kg	m³/kg	kJ/kg
230	0.91940	2928.8	0.76460	2926.9	0.57100	2922.9
240	0.93840	2949.0	0.78050	2947.2	0.58310	2943.5
250	0.95740	2969.2	0.79640	2967.5	0.59520	2964.1
260	0.97630	2989.4	0.81230	2987.8	0.60710	2984.6
270	0.99520	3009.6	0.82810	3008.1	0.61910	3005.2
280	1.01400	3029.8	0.84380	3028.5	0.63110	3025.7
290	1.03300	3050.0	0.85960	3048.8	0.64300	3046.2
300	1.05200	3070.4	0.87530	3069.2	0.65480	3066.7
310	1.07000	3090.7	0.89100	3089.6	0.66670	3087.2
320	1.08900	3111.0	0.90670	3110.0	0.67850	3107.7
330	0.10800	3131.5	0.92230	3130.4	0.69030	3128.3
340	0.12700	3151.9	0.93800	3150.9	0.70210	3148.9
350	0.14500	3172.4	0.95360	3171.4	0.71390	3169.5
360	0.16400	3192.9	0.96920	3192.0	0.72570	3190.1
370	0.18300	3213.5	0.98480	3212.7	0.73740	3210.9
380	0.20100	3234.2	1.00000	3233.4	0.74920	3231.7
390	0.22000	3254.9	1.01600	3254.1	0.76090	3252.5
400	0.23800	3275.6	1.03100	3274.9	0.77260	3273.3
410	0.25700	3296.4	1.04700	3295.7	0.78430	3294.2
420	0.27600	3317.3	1.06200	3316.6	0.79600	3315.2
430	0.29400	3338.2	1.07800	3337.5	0.80770	3336.2
440	0.31300	3359.2	1.09400	3358.5	0.81930	3357.2
450	0.33200	3380.2	1.10900	3379.6	0.83100	3378.3
	$p = 0.5$ MPa		$p = 0.6$ MPa		$p = 0.7$ MPa	
160	0.38360	2767.4	0.31660	2759.2	—	—
170	0.39420	2789.9	0.32580	2782.7	0.27680	2775.2
180	0.40460	2812.1	0.33470	2805.6	0.28470	2798.9
190	0.41480	2833.9	0.34340	2828.1	0.29240	2822.1
200	0.42490	2855.4	0.35210	2850.2	0.29990	2844.8
210	0.43490	2876.8	0.36060	2872.0	0.30740	2867.1
220	0.44490	2897.9	0.36900	2893.6	0.31470	2889.1
230	0.45480	2918.9	0.37740	2914.9	0.32200	2910.8
240	0.46460	2939.9	0.38570	2936.1	0.32920	2932.3
250	0.47440	2960.7	0.39390	2957.2	0.33630	2953.7
260	0.48410	2981.4	0.40210	2978.2	0.34340	2974.9
270	0.49380	3002.1	0.41020	2999.1	0.35050	2996.1
280	0.50340	3022.8	0.41830	3020.0	0.35750	3017.2
290	0.51300	3043.5	0.42640	3040.9	0.36440	3038.2
300	0.52260	3064.2	0.43440	3061.7	0.37140	3059.1

Table C.7 Specific volumes and specific enthalpies of steam at various temperatures and pressures—cont'd

T	v	h	v	h	v	h
°C	m³/kg	kJ/kg	m³/kg	kJ/kg	m³/kg	kJ/kg
310	0.53210	3084.8	0.44240	3082.4	0.37830	3080.0
320	0.54160	3105.5	0.45040	3103.2	0.38520	3100.9
330	0.55110	3126.1	0.45840	3124.0	0.0.3921	3121.8
340	0.56060	3146.8	0.46630	3144.8	0.39890	3142.7
350	0.57010	3167.5	0.47420	3165.6	0.40570	3163.6
360	0.57960	3188.3	0.48210	3186.4	0.41260	3184.6
370	0.58900	3209.1	0.49000	3207.4	0.41940	3205.6
380	0.59840	3230.0	0.49790	3228.4	0.42620	3226.6
390	0.60780	3250.9	0.50580	3249.3	0.43290	3247.7
400	0.61720	3271.8	0.51370	3270.3	0.43970	3268.7
410	0.62660	3292.7	0.52150	3291.3	0.44650	3289.8
420	0.63600	3313.7	0.52940	3312.4	0.45320	3310.9
430	0.64540	3334.8	0.53720	3333.5	0.46000	3332.1
440	0.65480	3355.9	0.54510	3354.6	0.46670	3353.3
450	0.66410	3377.0	0.55290	3375.8	0.47340	3374.5
460	0.67350	3398.2	0.56070	3397.1	0.48010	3395.8
470	0.68280	3419.5	0.56850	3418.4	0.48690	3417.2
480	0.69220	3440.8	0.57630	3439.7	0.49360	3438.6
490	0.70150	3462.2	0.58410	3461.1	0.50030	3460.0
500	0.71090	3483.6	0.59190	3482.6	0.50700	3481.5
510	0.72020	3505.1	0.59970	3504.1	0.51370	3503.1
520	0.72950	3526.7	0.60750	3525.7	0.52040	3524.7
530	0.73880	3548.3	0.61530	3547.4	0.52700	3546.4
540	0.74820	3570.0	0.62310	3569.1	0.53370	3568.1
550	0.75750	3591.7	0.63090	3590.8	0.54040	3589.9
560	—	—	—	—	0.54710	3611.8
570	—	—	—	—	0.55370	3633.7
580	—	—	—	—	0.56040	3655.7
590	—	—	—	—	0.56710	3677.7
600	—	—	—	—	0.57370	3699.8
610	—	—	—	—	0.58040	3722.0
620	—	—	—	—	0.58700	3744.2
630	—	—	—	—	0.59370	3766.4
640	—	—	—	—	0.60040	3788.8
650	—	—	—	—	0.60700	3811.2
	$p = 0.8$ MPa		$p = 0.9$ MPa		$p = 1.0$ MPa	
180	0.24710	2792.0	0.21780	2784.8	0.19440	2777.3
190	0.25400	2815.9	0.22420	2809.5	0.20020	2802.9

Continued

Table C.7 Specific volumes and specific enthalpies of steam at various temperatures and pressures—cont'd

T	υ	h	υ	h	υ	h
°C	m³/kg	kJ/kg	m³/kg	kJ/kg	m³/kg	kJ/kg
200	0.26080	2839.2	0.23040	2833.5	0.20590	2827.5
210	0.26750	2862.0	0.23640	2856.9	0.21150	2851.5
220	0.27400	2884.5	0.24230	2879.8	0.21690	2874.9
230	0.28050	2906.6	0.24810	2902.3	0.22230	2897.9
240	0.28690	2928.4	0.25390	2924.5	0.22750	2920.5
250	0.29320	2950.1	0.25960	2946.5	0.23270	2942.8
260	0.29950	2971.6	0.26520	2968.2	0.23780	2964.8
270	0.30570	2993.0	0.27080	2989.8	0.24290	2986.7
280	0.31190	3014.2	0.27640	3011.3	0.24800	3008.3
290	0.31800	3035.4	0.28190	3032.7	0.25300	3029.9
300	0.32410	3056.5	0.28740	3054.0	0.25800	3051.3
310	0.33020	3077.6	0.29280	3075.2	0.26290	3072.7
320	0.33630	3098.7	0.29830	3096.3	0.26780	3094.0
330	0.34230	3119.7	0.30370	3117.5	0.27270	3115.3
340	0.34840	3140.7	0.30910	3138.6	0.27760	3136.5
350	0.35440	3161.7	0.31440	3159.7	0.28250	3157.7
360	0.36040	3182.7	0.31980	3180.8	0.28730	3178.9
370	0.36640	3203.3	0.32510	3202.0	0.29210	3200.2
380	0.37230	3225.0	0.33040	3223.2	0.29700	3221.5
390	0.37830	3246.1	0.33580	3244.4	0.30180	3242.8
400	0.38420	3267.2	0.34110	3265.6	0.30660	3264.0
410	0.39020	3288.3	0.34640	3286.8	0.31130	3285.3
420	0.39610	3309.5	0.35170	3308.1	0.31610	3306.6
430	0.40200	3330.7	0.35690	3329.3	0.32090	3327.9
440	0.40790	3352.0	0.36220	3350.6	0.32560	3349.3
460	0.41970	3394.6	0.37270	3393.4	0.33510	3392.1
470	0.42560	3416.0	0.37800	3414.8	0.33990	3413.6
480	0.43150	3437.4	0.38320	3436.3	0.34460	3435.1
490	0.43740	3458.9	0.38850	3457.8	0.34930	3456.7
500	0.44320	3480.5	0.39370	3479.4	0.35400	3478.3
510	0.44910	3502.1	0.39890	3501.0	0.35880	3500.0
520	0.45500	3523.7	0.40410	3522.7	0.36350	3521.7
530	0.46080	3545.4	0.40940	3544.4	0.36820	3543.5
540	0.46670	3567.2	0.41460	3566.2	0.37290	3565.3
550	0.47250	3589.0	0.41980	3588.1	0.37760	3587.2
560	0.47840	3610.9	0.42500	3610.0	0.38230	3609.1
570	0.48420	3632.8	0.43020	3632.0	0.38700	3631.1
580	0.49010	3654.8	0.43540	3654.0	0.39160	3653.2
590	0.49590	3676.9	0.44060	3676.1	0.39630	3675.3

Table C.7 Specific volumes and specific enthalpies of steam at various temperatures and pressures—cont'd

T	υ	h	υ	h	υ	h
°C	m³/kg	kJ/kg	m³/kg	kJ/kg	m³/kg	kJ/kg
600	0.50180	3699.0	0.44580	3698.2	0.40100	3697.4
610	0.50760	3721.2	0.45100	3720.4	0.40570	3719.6
620	0.51340	3743.4	0.45620	3742.7	0.41040	3741.9
630	0.51930	3765.7	0.46140	3765.0	0.41510	3764.2
640	0.52510	3788.1	0.46660	3787.4	0.41970	3786.6
650	0.53090	3810.5	0.47180	3809.8	0.42440	3809.1
	$p = 1.1$ MPa		$p = 1.2$ MPa		$p = 1.3$ MPa	
190	0.18060	2796.0	0.16420	2788.9	—	—
200	0.18590	2821.5	0.16920	2815.2	0.15510	2808.8
210	0.19110	2846.1	0.17410	2840.5	0.15970	2834.8
220	0.19620	2870.0	0.17880	2865.0	0.16410	2859.9
230	0.20110	2893.4	0.18340	2888.9	0.16840	2884.2
240	0.20590	2916.4	0.18790	2912.2	0.17270	2908.0
250	0.21070	2939.0	0.19240	2935.2	0.17680	2931.3
260	0.21540	2961.2	0.19680	2957.8	0.18090	2954.2
270	0.22010	2983.4	0.20110	2980.2	0.18500	2976.8
280	0.22470	3005.3	0.20540	3002.3	0.18900	2999.2
290	0.22930	3027.1	0.20960	3024.2	0.19290	3021.4
300	0.23390	3048.7	0.21380	3046.0	0.19680	3043.4
310	0.23840	3070.2	0.21800	3067.7	0.20070	3065.2
320	0.24290	3091.6	0.22220	3089.3	0.20460	3086.9
330	0.24740	3113.0	0.22630	3110.8	0.20840	3108.5
340	0.25190	3134.4	0.23040	3132.2	0.21230	3130.1
350	0.25530	3155.7	0.23450	3153.7	0.21610	3151.7
360	0.26070	3177.0	0.23860	31.75.1	0.21990	3173.2
370	0.26520	3198.4	0.24260	3196.6	0.22360	3194.8
380	0.26960	3219.8	0.24670	3218.1	0.22740	3216.3
390	0.27390	3241.1	0.25070	3239.5	0.23110	3237.6
400	0.27830	3262.5	0.25480	3260.9	0.23480	3259.3
410	0.28270	3283.8	0.25880	3282.3	0.23860	3280.8
420	0.28700	3305.2	0.26280	3303.7	0.24230	3302.3
430	0.29140	3326.5	0.26680	3325.1	0.24600	3323.7
440	0.29570	3347.9	0.27080	3346.6	0.24970	3345.3
450	0.30000	3369.4	0.27480	3368.1	0.25340	3366.8
460	0.30440	3390.9	0.27870	3389.6	0.25700	3388.4
470	0.30870	3412.4	0.28270	3411.2	0.26070	3410.0
480	0.31300	3434.0	0.28670	3432.8	0.26440	3431.6
490	0.31730	3455.6	0.29060	3454.4	0.26800	3453.3

Continued

Table C.7 Specific volumes and specific enthalpies of steam at various temperatures and pressures—cont'd

T	υ	h	υ	h	υ	h
°C	m³/kg	kJ/kg	m³/kg	kJ/kg	m³/kg	kJ/kg
500	0.32160	3477.2	0.29460	3476.1	0.27170	3475.1
510	0.32590	3498.9	0.29850	3497.9	0.27530	3496.8
520	0.33020	3520.7	0.30250	3519.7	0.27900	3518.7
530	0.33450	3542.7	0.30640	3541.5	0.28260	3540.5
540	0.33880	3564.3	0.31030	3563.4	0.28630	3562.4
550	0.34300	3586.2	0.31420	3585.3	0.28990	3584.4
560	0.34730	3608.2	0.31820	3607.3	0.29350	3606.4
570	0.35160	3630.2	0.32210	3629.4	0.29720	3628.5
580	0.35590	3652.3	0.32600	3651.5	0.30080	3650.6
590	0.36010	3674.4	0.32990	3673.6	0.30440	3672.8
600	0.36440	3696.6	0.33390	3695.8	0.30800	3695.0
610	0.36860	3718.9	0.33780	3718.1	0.31160	3717.3
620	0.37290	3741.2	0.34170	3740.4	0.31530	3739.7
630	0.37720	3763.5	0.34560	3762.8	0.31890	3762.1
640	0.38140	3785.9	0.34950	3785.2	0.32250	3784.5
650	0.38570	3808.4	0.35340	3807.7	0.32610	3807.0
	$p = 1.4$ MPa		$p = 1.5$ MPa		$p = 1.6$ MPa	
200	0.14290	2802.1	0.13240	2795.3	—	—
210	0.14730	2828.9	0.13660	2822.9	0.12710	2816.7
220	0.15150	2854.5	0.14060	2849.2	0.13100	2843.7
230	0.15560	2879.5	0.14450	2874.7	0.13470	2869.8
240	0.15960	2903.7	0.14830	2899.3	0.13830	2894.9
250	0.16350	2927.4	0.15200	2923.4	0.14190	2919.4
260	0.16740	2950.6	0.15560	2947.0	0.14530	2943.3
270	0.17120	2973.5	0.15920	2970.2	0.14870	2966.7
280	0.17490	2996.0	0.16270	2993.0	0.15210	2989.8
290	0.17860	3018.0	0.16620	3015.6	0.15540	3012.7
300	0.18230	3040.7	0.16970	3037.9	0.15860	3035.2
310	0.18590	3062.7	0.17310	3060.0	0.16190	3057.5
320	0.18950	3084.5	0.17650	3082.1	0.16510	3079.7
330	0.19310	3106.3	0.17990	3104.0	0.16830	3101.7
340	0.19670	3128.0	0.18320	3125.8	0.17140	3123.6
350	0.20020	3149.6	0.18660	3147.6	0.17460	3145.5
360	0.20380	3171.2	0.18990	3169.3	0.17770	3167.4
370	0.20730	3192.9	0.19320	3191.1	0.18080	3189.3
380	0.21080	3214.6	0.19640	3212.8	0.18390	3211.1
390	0.21430	3236.2	0.19970	3234.5	0.18700	3232.8
400	0.21780	3257.7	0.20300	3256.1	0.19000	3254.5
410	0.22120	3279.3	0.20620	3277.7	0.19310	3276.2

Table C.7 Specific volumes and specific enthalpies of steam at various temperatures and pressures—cont'd

T	υ	h	υ	h	υ	h
°C	m³/kg	kJ/kg	m³/kg	kJ/kg	m³/kg	kJ/kg
420	0.22470	3300.8	0.20950	3299.3	0.19610	3297.9
430	0.22810	3322.3	0.21270	3320.9	0.19920	3319.5
440	0.23160	3343.9	0.21590	3342.6	0.20220	3341.2
450	0.23500	3365.5	0.21910	3364.2	0.20520	3362.9
460	0.23840	3387.1	0.22230	3385.9	0.20820	3384.6
470	0.24190	3408.8	0.22550	3407.6	0.21120	3406.4
480	0.24530	3430.5	0.22870	3429.3	0.21420	3428.1
490	0.24870	3452.2	0.23190	3451.1	0.21720	3449.9
500	0.25210	3474.0	0.23510	3472.9	0.22020	3471.8
510	0.25550	3495.8	0.23830	3494.7	0.22320	3493.7
520	0.25890	3517.6	0.24140	3516.6	0.22620	3515.6
530	0.26230	3539.5	0.24460	3538.6	0.22920	3537.6
540	0.26570	3561.5	0.24780	3560.5	0.23220	3559.6
550	0.26900	3583.5	0.25090	3582.5	0.23510	3581.6
560	0.27240	3605.5	0.25410	3604.6	0.23810	3603.7
570	0.27580	3627.6	0.25720	3626.8	0.24100	3625.9
580	0.27920	3649.8	0.26040	3649.0	0.24400	3648.1
590	0.28250	3672.0	0.26360	3671.2	0.24700	3670.4
600	0.28590	3694.3	0.26670	3693.5	0.24990	3692.7
610	0.28920	3716.6	0.26980	3715.8	0.25290	3715.0
620	0.29260	3738.9	0.27300	3738.2	0.25580	3737.4
630	0.29600	3761.3	0.27610	3760.6	0.25880	3759.8
640	0.29930	3783.8	0.27930	3783.1	0.26170	3782.3
650	0.30270	3806.3	0.28240	3805.6	0.26460	3804.9
	$p = 1.7$ MPa		$p = 1.8$ MPa		$p = 1.9$ MPa	
210	0.11880	2810.3	0.11140	2803.7	0.10470	2797.0
220	0.12250	2838.1	0.11500	2832.3	0.10820	2826.4
230	0.12610	2864.7	0.11840	2859.6	0.11150	2854.3
240	0.12960	2890.4	0.12180	2885.8	0.11480	2881.1
250	0.13290	2915.2	0.12500	2911.1	0.11790	2906.8
260	0.13620	2939.5	0.12810	2935.7	0.12090	2931.8
270	0.13950	2963.3	0.13120	2959.8	0.12390	2956.3
280	0.14270	2986.7	0.13430	2983.4	0.12680	2980.2
290	0.14580	3009.7	0.13730	3006.7	0.12960	3003.7
300	0.14890	3032.4	0.14020	3029.6	0.13250	3026.8
310	0.15200	3054.9	0.14310	3052.3	0.13530	3049.7
320	0.15500	3077.2	0.14600	3074.8	0.13800	3072.3
330	0.15800	3099.4	0.14890	3097.1	0.14070	3094.8

Continued

Table C.7 Specific volumes and specific enthalpies of steam at various temperatures and pressures—cont'd

T	v	h	v	h	v	h
°C	m³/kg	kJ/kg	m³/kg	kJ/kg	m³/kg	kJ/kg
340	0.16100	3121.5	0.15170	3119.3	0.14340	3117.1
350	0.16400	3143.5	0.15460	3141.4	0.14610	3139.3
360	0.16690	3165.4	0.15740	3163.5	0.14880	3161.5
370	0.16990	3187.4	0.16020	3185.6	0.15150	3183.7
380	0.17280	3209.3	0.16290	3207.6	0.15410	3205.8
390	0.17570	3231.2	0.16570	3229.5	0.15670	3227.8
400	0.17860	3253.0	0.16840	3251.3	0.15930	3249.7
410	0.18150	3274.7	0.17120	3273.1	0.16190	3271.6
420	0.18440	3296.4	0.17390	3294.9	0.16450	3293.4
430	0.18720	3318.1	0.17660	3316.7	0.16710	3315.3
440	0.19010	3339.8	0.17930	3338.5	0.16970	3337.1
450	0.19290	3361.6	0.18200	3360.3	0.17230	3359.0
460	0.19580	3383.4	0.18470	3382.1	0.17480	3380.8
470	0.19860	3405.2	0.18740	3403.9	0.17740	3402.7
480	0.20150	3427.0	0.19010	3425.8	0.17990	3424.6
490	0.20430	3448.8	0.19280	3447.7	0.18250	3446.6
500	0.20710	3470.7	0.19540	3469.6	0.18500	3468.5
510	0.20990	3492.6	0.19810	3491.6	0.18750	3490.5
520	0.21270	3514.6	0.20080	3513.6	0.19010	3512.5
530	0.21550	3536.6	0.20340	3535.6	0.19260	3534.6
540	0.21830	3558.6	0.20610	3557.7	0.19510	3556.7
550	0.22110	3580.7	0.20870	3579.8	0.19760	3578.9
560	0.22390	3602.8	0.21140	3602.0	0.20010	3601.1
570	0.22670	3625.0	0.21400	3624.2	0.20260	3623.3
580	0.22950	3647.3	0.21660	3646.4	0.20510	3645.6
590	0.23230	3669.3	0.21930	3668.7	0.20760	3667.9
600	0.23510	3691.9	0.22190	3691.1	0.21010	3690.3
610	0.23790	3714.2	0.22460	3713.5	0.21260	3712.7
620	0.24060	3736.6	0.22720	3735.9	0.21510	3735.1
630	0.24340	3759.1	0.22980	3758.4	0.21760	3757.6
640	0.24620	3781.6	0.23240	3780.9	0.22010	3780.2
650	0.24900	3804.2	0.23510	3803.5	0.22260	3802.8
	p = 2.0 MPa		*p* = 2.1 MPa		*p* = 2.2 MPa	
220	0.10210	2820.4	0.09657	2814.2	0.09152	2807.8
230	0.10530	2849.0	0.09972	2843.5	0.09460	2837.9
240	0.10840	2876.3	0.10274	2871.4	0.09754	2866.4
250	0.11150	2902.5	0.10570	2898.2	0.10037	2893.7
260	0.11440	2927.9	0.10850	2924.0	0.10310	2919.9
270	0.11720	2952.7	0.11120	2949.0	0.10580	2945.4

Table C.7 Specific volumes and specific enthalpies of steam at various temperatures and pressures—cont'd

T	v	h	v	h	v	h
°C	m³/kg	kJ/kg	m³/kg	kJ/kg	m³/kg	kJ/kg
280	0.1200	2976.9	0.11390	2973.5	0.10840	2970.2
290	0.12280	3000.6	0.11660	2997.5	0.11090	2994.4
300	0.12550	3024.0	0.11920	3021.1	0.11340	3018.2
310	0.12820	3047.0	0.12170	3044.4	0.11590	3041.7
320	0.13080	3069.8	0.12430	3067.3	0.11830	3064.8
330	0.13340	3092.4	0.12680	3090.0	0.12070	3087.7
340	0.13600	3114.9	0.12930	3112.6	0.12310	3110.4
350	0.13860	3137.2	0.13170	3135.1	0.12550	3133.0
360	0.14110	3159.5	0.13420	3157.6	0.12780	3155.6
370	0.14360	3181.8	0.13660	3180.0	0.13010	3178.1
380	0.14620	3204.0	0.13900	3202.2	0.13250	3200.4
390	0.14870	3226.1	0.14140	3224.4	0.13480	3222.7
400	0.15120	3248.1	0.14380	3246.5	0.13700	3244.8
410	0.15360	3270.0	0.14610	3268.5	0.13930	3266.9
420	0.15610	3291.9	0.14850	3290.5	0.14160	3289.0
430	0.15860	3313.8	0.15080	3312.4	0.14380	3311.0
440	0.16100	3335.8	0.15320	3334.4	0.14610	3333.0
450	0.16350	3357.7	0.15550	3356.3	0.14830	3355.0
460	0.16590	3379.6	0.15790	3378.3	0.15050	3377.0
470	0.16840	3401.5	0.16020	3400.3	0.15280	3399.1
480	0.17080	3423.5	0.16250	3422.3	0.15500	3421.1
490	0.17320	3445.4	0.16480	3444.3	0.15720	3443.2
500	0.17560	3467.4	0.16710	3466.3	0.1594	3465.2
510	0.17800	3489.5	0.16940	3488.4	0.16160	3487.3
520	0.18040	3511.5	0.17170	3510.5	0.16380	3509.4
530	0.18280	3533.6	0.17400	3532.6	0.16600	3531.6
540	0.18520	3555.8	0.17630	3554.8	0.16820	3553.8
550	0.18760	3578.0	0.17860	3577.0	0.17030	3576.1
560	0.19000	3600.2	0.18080	3599.3	0.17250	3598.4
570	0.19240	3622.5	0.18310	3621.6	0.17470	3620.7
580	0.19480	3644.8	0.18540	3643.9	0.17690	3643.1
590	0.19720	3667.1	0.18760	3666.2	0.17900	3665.5
600	0.19950	3689.5	0.18990	3688.6	0.18120	3687.9
610	0.20190	3711.9	0.19220	3711.1	0.18340	3710.4
620	0.20430	3734.4	0.19450	3733.6	0.18550	3732.9
630	0.20660	3756.9	0.19670	3756.2	0.18770	3755.4
640	2090.00000	3779.5	0.19900	3778.8	0.18990	3778.0
650	0.21140	3802.1	0.20120	3801.4	0.19200	3800.7

Continued

Table C.7 Specific volumes and specific enthalpies of steam at various temperatures and pressures—cont'd

T	v	h	v	h	v	h
°C	m³/kg	kJ/kg	m³/kg	kJ/kg	m³/kg	kJ/kg
	$p = 2.3$ MPa		$p = 2.4$ MPa		$p = 2.5$ MPa	
220	0.08690	2801.2	—	—	—	—
230	0.08992	2832.2	0.08561	2826.4	0.08164	2820.4
240	0.09279	2861.4	0.08842	2856.2	0.08439	2851.0
250	0.09554	2889.2	0.09110	2884.6	0.08701	2879.9
260	0.09819	2915.8	0.09368	2911.7	0.08953	2907.5
270	0.10077	2941.6	0.09619	2937.9	0.09197	2934.1
280	0.10328	2966.7	0.09863	2963.3	0.09434	2959.8
290	0.10570	2991.2	0.10101	2988.1	0.09665	2984.9
300	0.10810	3015.3	0.10330	3012.4	0.09822	3009.4
310	0.11050	3039.0	0.10560	3036.2	0.10114	3033.4
320	0.11290	3062.3	0.10790	3059.7	0.10334	3057.1
330	0.11520	3085.3	0.11020	3082.9	0.10550	3080.5
340	0.11750	3108.2	0.11240	3105.9	0.10760	3103.6
350	0.11980	3130.9	0.11460	3128.8	0.10980	3126.6
360	0.12200	3153.6	0.11670	3151.6	0.11190	3149.6
370	0.12430	3176.2	0.11890	3174.3	0.11390	3172.4
380	0.12650	3198.6	0.12100	3196.8	0.11600	3195.0
390	0.12870	3221.0	0.12310	3219.2	0.1180	3217.5
400	0.13090	3243.2	0.12530	3241.5	0.12010	3239.9
410	0.13310	3265.4	0.12740	3263.8	0.12210	3262.2
420	0.13520	3287.5	0.12940	3286.0	0.12410	3284.5
430	0.13740	3309.6	0.13150	3308.1	0.12610	3306.7
440	0.13960	3331.6	0.13360	3330.2	0.12810	3328.9
450	0.14170	3353.7	0.13570	3352.4	0.13010	3351.0
460	0.14380	3375.8	0.13770	3374.5	0.13210	3373.2
470	0.14600	3397.8	0.13980	3396.6	0.13400	3395.4
480	0.14810	3419.9	0.14180	3418.7	0.13600	3417.5
490	0.15020	3442.0	0.14380	3440.9	0.13800	3439.7
500	0.15230	3464.1	0.14590	3463.0	0.13990	3461.9
510	0.15440	3486.3	0.14790	3485.2	0.14190	3484.1
520	0.15660	3508.4	0.14990	3507.4	0.14380	3506.4
530	0.15870	3530.6	0.15190	3529.6	0.14580	3528.7
540	0.16080	3552.9	0.15400	3551.9	0.14770	3551.0
550	0.16280	3575.2	0.15600	3574.2	0.14960	3573.3
560	0.16490	3597.5	0.15800	3596.6	0.15160	3595.7
570	0.16700	3619.8	0.16000	3619.0	0.15350	3618.1
580	0.16910	3642.2	0.16200	3641.4	0.15540	3640.5
590	0.17120	3664.6	0.16400	3663.8	0.15730	3663.0

Table C.7 Specific volumes and specific enthalpies of steam at various temperatures and pressures—cont'd

T	υ	h	υ	h	υ	h
°C	m³/kg	kJ/kg	m³/kg	kJ/kg	m³/kg	kJ/kg
600	0.17320	3687.1	0.16600	3686.3	0.15920	3685.5
610	0.17530	3709.6	0.16790	3708.8	0.16120	3708.0
620	0.17740	3732.0	0.16990	3731.4	0.16310	3730.6
630	0.17950	3754.7	0.17190	3754.0	0.16500	3753.2
640	0.18150	3777.3	0.17390	3776.6	0.16690	3775.9
650	0.18360	3800.0	0.17590	3799.3	0.16880	3798.6
	$p = 2.6$ MPa		$p = 2.7$ MPa		$p = 2.8$ MPa	
230	0.07796	2814.2	0.07455	2808.0	—	—
240	0.08066	2845.6	0.07721	2840.1	0.07399	2834.5
250	0.08323	2875.1	0.07973	2870.3	0.07647	2865.3
260	0.08570	2903.2	0.08214	2898.9	0.07883	2894.5
270	0.08807	2930.2	0.08445	2926.3	0.08110	2922.3
280	0.09037	2956.3	0.08670	2952.7	0.08329	2949.1
290	0.09262	2981.6	0.08889	2978.4	0.08542	2975.1
300	0.09482	3006.4	0.09103	3003.4	0.08751	3000.4
310	0.09698	3030.7	0.09313	3027.9	0.08955	3025.1
320	0.09911	3054.6	0.09519	3051.9	0.09155	3049.3
330	0.10120	3078.1	0.09722	3075.6	0.09353	3073.2
340	0.10327	3101.3	0.09923	3099.0	0.09547	3096.7
350	0.10530	3124.5	0.10121	3122.3	0.09740	3120.1
360	0.10730	3147.5	0.10318	3145.5	0.09930	3143.4
370	0.10930	3170.4	0.10510	3168.5	0.10118	3166.6
380	0.11130	3193.2	0.10700	3191.3	0.10305	3189.5
390	0.11330	3215.8	0.10890	3214.0	0.10490	3212.3
400	0.11530	3238.3	0.11080	3236.6	0.10670	3234.9
410	0.11720	3260.6	0.11270	3259.0	0.10860	3257.5
420	0.11920	3283.0	0.11460	3281.4	0.11040	3279.9
430	0.12110	3305.2	0.11650	3303.8	0.11220	3302.3
440	0.12300	3327.5	0.11830	3326.1	0.11400	3324.7
450	0.12500	3349.7	0.12020	3348.4	0.11580	3347.0
460	0.12690	3371.9	0.12200	3370.6	0.11760	3369.4
470	0.12880	3394.1	0.12390	3392.9	0.11930	3391.7
480	0.13070	3416.4	0.12570	3415.2	0.12110	3414.0
490	0.13260	3438.6	0.12750	3437.4	0.12290	3436.3
500	0.13440	3460.8	0.12940	3459.7	0.12460	3458.6
510	0.13630	3483.1	0.13120	3482.0	0.12640	3480.9
520	0.13820	3505.4	0.13300	3504.3	0.12810	3503.3
530	0.14010	3527.7	0.13480	3526.7	0.12990	3525.7

Continued

Table C.7 Specific volumes and specific enthalpies of steam at various temperatures and pressures—cont'd

T	v	h	v	h	v	h
°C	m³/kg	kJ/kg	m³/kg	kJ/kg	m³/kg	kJ/kg
540	0.14190	3550.0	0.13660	3549.0	0.13160	3548.1
550	0.14380	3572.4	0.13840	3571.4	0.13330	3570.5
560	0.14560	3594.8	0.14020	3593.8	0.13510	3592.9
570	0.14750	3617.2	0.14200	3616.3	0.13680	3615.4
580	0.14930	3639.7	0.14370	3638.8	0.13850	3638.0
590	0.15120	3662.2	0.14560	3661.3	0.14020	3660.5
600	0.15300	3684.7	0.14730	3683.9	0.14200	3683.1
610	0.15490	3707.2	0.14910	3706.5	0.14370	3705.7
620	0.15670	3729.9	0.15080	3729.1	0.14540	3728.3
630	0.15860	3752.5	0.15260	3751.8	0.14710	3751.0
640	0.16040	3775.2	0.15440	3774.5	0.14880	3773.8
650	0.16220	3797.9	0.15620	3797.2	0.15050	3796.5
	$p = 2.9$ MPa		$p = 3.0$ MPa		$p = 3.1$ MPa	
240	0.07099	2828.8	0.06818	2823.0	0.06554	2817.0
250	0.07343	2860.3	0.07058	2855.2	0.06791	2850.0
260	0.07574	2890.0	0.07286	2885.5	0.07016	2880.8
270	0.07796	2918.3	0.07504	2914.2	0.07230	2910.0
280	0.08011	2945.4	0.07714	2941.8	0.07436	2938.0
290	0.08219	2971.7	0.07918	2968.4	0.07635	2964.9
300	0.08422	2997.3	0.08116	2994.2	0.07829	2991.0
310	0.08621	3022.2	0.08310	3019.3	0.08018	3016.4
320	0.08816	3046.7	0.08500	3044.0	0.08204	3041.3
330	0.09008	3070.7	0.08687	3068.2	0.08386	3065.7
340	0.09198	3094.4	0.08871	3092.1	0.08565	3089.7
350	0.09385	3118.0	0.09053	3115.7	0.08742	3113.5
360	0.09569	3141.4	0.09232	3139.3	0.08917	3137.3
370	0.09752	3164.6	0.09410	3162.7	0.09090	3160.7
380	0.09933	3187.7	0.09586	3185.8	0.09261	3184.0
390	0.10113	3210.5	0.09760	3208.8	0.09432	3207.0
400	0.10291	3233.3	0.09933	3231.6	0.09599	3229.9
410	0.10470	3255.9	0.10105	3254.3	0.09766	3252.7
420	0.10640	3278.4	0.10276	3276.9	0.09932	3275.3
430	0.10820	3300.9	0.10450	3299.4	0.10097	3297.9
440	0.10990	3323.3	0.10610	3321.9	0.10261	3320.5
450	0.11170	3345.7	0.10780	3344.4	0.10420	3343.0
460	0.11340	3368.0	0.10950	3366.8	0.10590	3365.5
470	0.11510	3390.4	0.11120	3389.2	0.10750	3387.9
480	0.11680	3412.8	0.11280	3411.6	0.10910	3410.4
490	0.11850	3435.1	0.11450	3434.0	0.11070	3432.8

Table C.7 Specific volumes and specific enthalpies of steam at various temperatures and pressures—cont'd

T	v	h	v	h	v	h
°C	m³/kg	kJ/kg	m³/kg	kJ/kg	m³/kg	kJ/kg
500	0.12020	3457.5	0.11610	3456.4	0.11230	3455.3
510	0.12190	3479.9	0.11780	3478.8	0.11390	3477.7
520	0.12360	3502.3	0.11940	3501.2	0.11550	3500.2
530	0.12530	3524.7	0.12100	3523.7	0.11710	3522.7
540	0.12700	3547.1	0.12270	3546.1	0.11860	3545.2
550	0.12870	3569.5	0.12430	3568.6	0.12020	3567.7
560	0.13030	3592.0	0.12590	3591.1	0.12180	3590.2
570	0.13200	3614.6	0.12750	3613.7	0.12340	3612.8
580	0.13360	3637.1	0.12920	3636.3	0.12490	3635.4
590	0.13530	3659.7	0.13080	3658.9	0.12650	3658.0
600	0.13700	3682.3	0.13240	3681.5	0.12800	3680.7
610	0.13870	3704.9	0.13400	3704.1	0.12960	3703.4
620	0.14030	3727.6	0.13560	3726.8	0.13120	3726.1
630	0.14200	3750.3	0.13720	3749.6	0.13270	3748.8
640	0.14360	3773.0	0.13880	3772.3	0.13420	3771.6
650	0.14530	3795.8	0.14040	3795.1	0.13580	3794.4
	$p = 3.2$ MPa		$p = 3.3$ MPa		$p = 3.4$ MPa	
240	0.06306	2810.9	0.06072	2804.7	—	—
250	0.06540	2844.7	0.06305	2839.3	0.06082	2833.8
260	0.06762	2876.2	0.06523	2871.4	0.06298	2866.5
270	0.06973	2905.9	0.06730	2901.6	0.06502	2897.3
280	0.07175	2934.2	0.06929	2930.4	0.06698	2926.5
290	0.07370	2961.5	0.07121	2958.0	0.06886	2954.5
300	0.07560	2987.9	0.07307	2984.7	0.07068	2981.5
310	0.07745	3013.5	0.07488	3010.6	0.07246	3007.6
320	0.07926	3038.6	0.07665	3035.9	0.07419	3033.1
330	0.08104	3063.2	0.07839	3060.6	0.07589	3058.0
340	0.08279	3087.4	0.08010	3085.0	0.07756	3082.6
350	0.08452	3111.3	0.08178	3109.1	0.07920	3106.8
360	0.08622	3135.2	0.08344	3133.2	0.08083	3131.0
370	0.08790	3158.7	0.08508	3156.8	0.08243	3154.8
380	0.08956	3182.1	0.08670	3180.2	0.08401	3178.3
390	0.09121	3205.2	0.08831	3203.5	0.08557	3201.7
400	0.09280	3228.2	0.08990	3226.5	0.08713	3224.8
410	0.09447	3251.1	0.09148	3249.5	0.08867	3247.8
420	0.09609	3273.8	0.09305	3272.3	0.09020	3270.7
430	0.09769	3296.5	0.09461	3295.0	0.09172	3293.5
440	0.09929	3319.1	0.09617	3317.7	0.09323	3316.3

Continued

Table C.7 Specific volumes and specific enthalpies of steam at various temperatures and pressures—cont'd

T	v	h	v	h	v	h
°C	m³/kg	kJ/kg	m³/kg	kJ/kg	m³/kg	kJ/kg
450	0.10087	3341.6	0.09771	3340.3	0.09473	3338.9
460	0.10245	3364.2	0.09924	3362.9	0.09622	3361.6
470	0.10400	3386.7	0.10077	3385.4	0.09771	3384.2
480	0.10560	3409.2	0.10229	3408.0	0.09919	3406.8
490	0.10710	3431.7	0.10380	3430.5	0.10067	3429.4
500	0.10870	3454.2	0.10530	3453.0	0.10213	3451.9
510	0.11020	3476.6	0.10680	3475.6	0.10360	3474.5
520	0.11180	3499.1	0.10830	3498.1	0.10510	3497.1
530	0.11330	3521.6	0.10980	3520.6	0.10650	3519.6
540	0.11490	3544.2	0.11130	3543.2	0.10800	3542.2
550	0.11640	3566.7	0.11280	3565.8	0.10940	3564.8
560	0.11790	3589.3	0.11430	3588.4	0.11080	3587.5
570	0.11940	3611.9	0.11570	3611.0	0.11230	3610.2
580	0.12090	3634.6	0.11720	3633.7	0.11370	3632.8
590	0.12250	3657.2	0.11870	3656.4	0.11510	3655.6
600	0.12400	3679.9	0.12020	3679.1	0.11660	3678.3
610	0.12550	3702.6	0.12160	3701.8	0.11800	3701.0
620	0.12700	3725.3	0.12310	3724.6	0.11940	3723.8
630	0.12850	3748.1	0.12450	3747.3	0.12080	3746.6
640	0.13000	3770.9	0.12600	3770.2	0.12220	3769.4
650	0.13150	3793.7	0.12750	3793.0	0.12370	3792.3
	$p = 3.5$ MPa		$p = 3.6$ MPa		$p = 3.7$ MPa	
250	0.05871	2828.1	0.05671	2822.4	0.05482	2816.5
260	0.06085	2861.6	0.05883	2856.6	0.05692	2851.5
270	0.06287	2892.9	0.06083	2888.5	0.05890	2884.0
280	0.06479	2922.6	0.06273	2918.6	0.06077	2914.6
290	0.06664	2950.9	0.06455	2947.3	0.06256	2943.7
300	0.06843	2978.2	0.06631	2974.9	0.06429	2971.6
310	0.07017	3004.6	0.06802	3001.6	0.06597	2998.6
320	0.07187	3030.3	0.06968	3027.6	0.06761	3024.7
330	0.07354	3055.5	0.07131	3052.9	0.06920	3050.3
340	0.07517	3080.2	0.07291	3077.7	0.07077	3075.3
350	0.07678	3104.6	0.07448	3102.3	0.07231	3100.0
360	0.07836	3128.9	0.07603	3126.8	0.07383	3124.6
370	0.07992	3152.8	0.07756	3150.8	0.07532	3148.8
380	0.08146	3176.4	0.07907	3174.6	0.07680	3172.7
390	0.08300	3199.9	0.08056	3198.1	0.07826	3196.9
400	0.08451	3223.1	0.08204	3221.4	0.07970	3219.7
410	0.08602	3246.2	0.08351	3244.6	0.08114	3243.0

Table C.7 Specific volumes and specific enthalpies of steam at various temperatures and pressures—cont'd

T	υ	h	υ	h	υ	h
°C	m³/kg	kJ/kg	m³/kg	kJ/kg	m³/kg	kJ/kg
420	0.08751	3269.2	0.08496	3267.7	0.08256	3266.1
430	0.08899	3292.0	0.08641	3290.6	0.08397	3389.1
440	0.09046	3314.8	0.08784	3313.4	0.08537	3312.9
450	0.09192	3337.6	0.08927	3336.2	0.08676	3334.4
460	0.09338	3360.3	0.09069	3359.0	0.08815	3357.7
470	0.09483	3382.9	0.09210	3381.7	0.08953	3380.4
480	0.09627	3405.6	0.09351	3404.4	0.09090	3403.2
490	0.09770	3428.2	0.09491	3427.0	0.09226	3425.9
500	0.09913	3450.8	0.09630	3449.7	0.09362	3448.6
510	0.10056	3473.4	0.09769	3472.3	0.09497	3471.2
520	0.10198	3496.0	0.09907	3495.0	0.09632	3493.9
530	0.10340	3518.6	0.10045	3517.6	0.09767	3516.6
540	0.10480	3541.3	0.10182	3540.3	0.09901	3539.3
550	0.10620	3563.9	0.10320	3563.0	0.10034	3562.0
560	0.10760	3586.6	0.10460	3585.7	0.10167	3584.8
570	0.10900	3609.3	0.10590	3608.4	0.10300	3607.5
580	0.11040	3632.0	0.10730	3631.1	0.10430	3630.3
590	0.11180	3654.7	0.10860	3653.9	0.10560	3653.1
600	0.11320	3677.5	0.11000	3676.7	0.10700	3675.9
610	0.11460	3700.2	0.11130	3699.4	0.10830	3698.7
620	0.11600	3723.0	0.11270	3722.3	0.10960	3721.5
630	0.11730	3745.8	0.11400	3745.1	0.11090	3744.4
640	0.11870	3768.7	0.11540	3768.0	0.11220	3767.3
650	0.12010	3791.6	0.11670	3790.9	0.11350	3790.2
	$p = 3.8$ MPa		$p = 3.9$ MPa		$p = 4.0$ MPa	
250	0.05302	2810.5	0.05130	2804.4	—	—
260	0.05511	2846.3	0.05338	2841.0	0.05174	2835.6
270	0.05706	2879.4	0.05532	2874.8	0.05366	2870.1
280	0.05891	2910.5	0.05715	2906.4	0.05547	2902.2
290	0.06068	2940.0	0.05889	2936.3	0.05719	2932.5
300	0.06238	2968.3	0.06057	2964.9	0.05885	2961.5
310	0.06403	2995.5	0.06220	2992.4	0.06045	2989.2
320	0.06564	3021.9	0.06377	3019.1	0.06200	3016.2
330	0.06721	3047.6	0.06531	3045.0	0.06351	3042.3
340	0.06875	3072.8	0.06682	3070.4	0.06499	3067.9
350	0.07026	3097.7	0.06830	3095.4	0.06645	3093.1
360	0.07174	3122.5	0.06976	3120.0	0.06787	3118.2
370	0.07320	3146.7	0.07119	3144.7	0.06928	3142.7

Continued

Table C.7 Specific volumes and specific enthalpies of steam at various temperatures and pressures—cont'd

T	υ	h	υ	h	υ	h
°C	m³/kg	kJ/kg	m³/kg	kJ/kg	m³/kg	kJ/kg
380	0.07465	3170.7	0.07260	3168.8	0.07066	3166.9
390	0.07670	3195.5	0.07400	3192.7	0.07203	3190.8
400	0.07749	3218.0	0.07538	3216.3	0.07339	3214.5
410	0.07889	3241.3	0.07676	3239.7	0.07473	3238.0
420	0.08028	3264.5	0.07811	3262.9	0.07606	3261.4
430	0.08166	3287.6	0.07946	3286.1	0.07738	3284.6
440	0.08303	3310.6	0.08080	3309.1	0.07869	3307.7
450	0.08439	3333.5	0.08213	3332.1	0.07999	3330.7
460	0.08574	3356.3	0.08345	3355.0	0.08128	3353.7
470	0.08708	3379.1	0.08477	3377.9	0.08257	3376.6
480	0.08842	3401.9	0.08607	3400.7	0.08384	3399.5
490	0.08975	3424.7	0.08737	3423.5	0.08512	3422.3
500	0.09108	3447.4	0.08867	3446.3	0.08638	3445.2
510	0.09240	3470.2	0.08996	3469.1	0.08764	3468.0
520	0.09372	3492.9	0.09125	3491.8	0.08890	3490.8
530	0.09503	3515.6	0.09253	3514.6	0.09015	3513.6
540	0.09634	3538.3	0.09380	3537.3	0.09140	3536.4
550	0.09764	3561.0	0.09507	3560.1	0.09264	3559.2
560	0.09893	3583.8	0.09634	3582.9	0.09387	3582.0
570	0.10023	3606.6	0.09760	3605.8	0.09510	3604.9
580	0.10152	3629.4	0.09886	3628.6	0.09634	3627.7
590	0.10280	3652.2	0.10012	3651.4	0.09756	3650.6
600	0.10410	3675.1	0.10137	3674.2	0.09879	3673.4
610	0.10540	3697.9	0.10260	3697.1	0.10001	3696.3
620	0.10660	3720.7	0.10390	3720.0	0.10123	3719.2
630	0.10790	3743.6	0.10510	3742.9	0.10245	3742.2
640	0.10920	3766.6	0.10640	3765.8	0.10370	3765.1
650	0.11050	3789.5	0.10760	3788.8	0.10490	3788.1
	$p = 4.1$ MPa		$p = 4.2$ MPa		$p = 4.3$ MPa	
260	0.05017	2830.2	0.04867	2824.6	0.04723	2818.9
270	0.05208	2865.3	0.05056	2860.4	0.04912	2855.5
280	0.05387	2898.0	0.05234	2893.7	0.05089	2889.3
290	0.05558	2928.7	0.05403	2924.9	0.05256	2921.0
300	0.05721	2958.0	0.05564	2954.6	0.05415	2951.0
310	0.05878	2986.2	0.05720	2983.0	0.05568	2979.8
320	0.60310	3013.3	0.05870	3010.4	0.05717	3007.4
330	0.06180	3039.6	0.06017	3036.9	0.05861	3034.2
340	0.06325	3065.4	0.06160	3062.9	0.06002	3060.4
350	0.06468	3090.8	0.06300	3088.4	0.06139	3086.1

Table C.7 Specific volumes and specific enthalpies of steam at various temperatures and pressures—cont'd

T	υ	h	υ	h	υ	h
°C	m³/kg	kJ/kg	m³/kg	kJ/kg	m³/kg	kJ/kg
360	0.06608	3116.0	0.06437	3113.8	0.06275	3111.6
370	0.06746	3140.6	0.06573	3138.6	0.06407	3136.5
380	0.06882	3165.0	0.06706	3163.0	0.06538	3161.1
390	0.07016	3189.0	0.06838	3187.2	0.06667	3185.3
400	0.07149	3212.8	0.06968	3211.1	0.06795	3209.3
410	0.07280	3236.4	0.07097	3234.7	0.06921	3233.1
420	0.07410	3259.8	0.07224	3258.2	0.07047	3256.6
430	0.07540	3283.1	0.07351	3281.6	0.07171	3280.1
440	0.07668	3306.3	0.07476	3304.8	0.07294	3303.4
450	0.07795	3329.4	0.07601	3328.0	0.07416	3326.6
460	0.07921	3352.4	0.07725	3351.1	0.07537	3349.7
470	0.08047	3375.4	0.07848	3374.1	0.07658	3372.8
480	0.08172	3398.3	0.07970	3397.1	0.07778	3395.8
490	0.08297	3421.2	0.08092	3420.0	0.07897	3418.8
500	0.08421	3444.0	0.08213	3442.9	0.08015	3441.8
510	0.08544	3466.9	0.08334	3465.8	0.08133	3464.7
520	0.08667	3489.7	0.08454	3488.7	0.08251	3487.6
530	0.08789	3512.6	0.08574	3511.5	0.08368	3510.5
540	0.08911	3535.4	0.08694	3534.4	0.08485	3533.4
550	0.09032	3558.2	0.08812	3557.2	0.08601	3556.3
560	0.09152	3581.1	0.08929	3580.2	0.08716	3579.2
570	0.09273	3604.0	0.09047	3603.1	0.08832	3602.2
580	0.09394	3626.9	0.09165	3626.0	0.08947	3625.1
590	0.09514	3649.7	0.09282	3648.9	0.09062	3648.1
600	0.09633	3672.6	0.09399	3671.8	0.09176	3671.0
610	0.09753	3695.5	0.09516	3694.8	0.09290	3694.0
620	0.09872	3718.4	0.09633	3717.7	0.09404	3716.9
630	0.09991	3741.4	0.09749	3740.7	0.09518	3739.9
640	0.10110	3764.4	0.09865	3763.7	0.09632	3762.9
650	0.10228	3787.4	0.09981	3786.7	0.09745	3786.0
	p = 4.4 MPa		*p* = 4.5 MPa		*p* = 4.6 MPa	
260	0.04585	2813.0	0.04454	2807.1	0.04327	2801.0
270	0.04774	2850.4	0.04641	2845.3	0.04514	2840.1
280	0.04949	2884.9	0.04816	2880.4	0.04688	2875.8
290	0.05115	2917.1	0.04980	2913.1	0.04850	2909.0
300	0.05272	2947.5	0.05136	2943.9	0.05005	2940.3
310	0.05424	2976.5	0.05285	2973.3	0.05153	2970.0
320	0.05570	3004.5	0.05430	3001.5	0.05295	2998.5

Continued

Table C.7 Specific volumes and specific enthalpies of steam at various temperatures and pressures—cont'd

T	v	h	v	h	v	h
°C	m³/kg	kJ/kg	m³/kg	kJ/kg	m³/kg	kJ/kg
330	0.05712	3031.5	0.05570	3028.7	0.05434	3026.0
340	0.05851	3057.8	0.05706	3055.3	0.05568	3052.7
350	0.05986	3083.7	0.05840	3081.3	0.05700	3078.9
360	0.06119	3109.4	0.05971	3107.2	0.05828	3105.0
370	0.06250	3134.4	0.06099	3132.3	0.05955	3130.2
380	0.06378	3159.1	0.06225	3157.1	0.06079	3155.2
390	0.06505	3183.5	0.06350	3181.6	0.06201	3179.8
400	0.06630	3207.6	0.06473	3205.8	0.06322	3204.0
410	0.06754	3231.4	0.06595	3229.7	0.06442	3228.0
420	0.06877	3255.1	0.06715	3253.5	0.06560	3251.9
430	0.06999	3278.6	0.06834	3277.0	0.06677	3275.5
440	0.07119	3301.9	0.06953	3300.5	0.06793	3299.0
450	0.07239	3325.2	0.07070	3323.8	0.06909	3322.4
460	0.07358	3348.4	0.07187	3347.1	0.07023	3345.7
470	0.07476	3371.5	0.07303	3370.3	0.07137	3369.0
480	0.07594	3394.6	0.07418	3393.4	0.07250	3392.2
490	0.07710	3417.6	0.07532	3416.5	0.07362	3415.3
500	0.07827	3440.6	0.07646	3439.5	0.07474	3438.4
510	0.07942	3463.6	0.07760	3462.5	0.07585	3461.4
520	0.08057	3486.8	0.07872	3485.5	0.07696	3484.4
530	0.08172	3509.5	0.07985	3508.5	0.07806	3507.5
540	0.08286	3532.4	0.08097	3531.4	0.07915	3530.5
550	0.08400	3555.4	0.08208	3554.4	0.08024	3553.5
560	0.08513	3578.5	0.08319	3577.4	0.08133	3576.5
570	0.08626	3601.3	0.08429	3600.4	0.08241	3599.5
580	0.08739	3624.3	0.08540	3623.4	0.08350	3622.6
590	0.08851	3647.2	0.08650	3646.4	0.08457	3645.6
600	0.08963	3670.2	0.08760	3669.4	0.08565	3668.6
610	0.09075	3693.2	0.08869	3692.4	0.08672	3691.6
620	0.09186	3716.2	0.08978	3715.4	0.08779	3714.6
630	0.09298	3739.2	0.09087	3738.4	0.08886	3737.7
640	0.09409	3762.2	0.09196	3761.5	0.08992	3760.8
650	0.09520	3785.3	0.09304	3784.6	0.09099	3783.9
	$p = 4.7$ MPa		$p = 4.8$ MPa		$p = 4.9$ MPa	
270	0.04392	2834.8	0.04274	2829.4	0.04162	2824.0
280	0.04565	2871.2	0.04447	2866.6	0.04333	2861.8
290	0.04726	2904.9	0.04607	2900.8	0.04493	2896.6
300	0.04879	2936.6	0.04759	2932.9	0.04644	2929.2
310	0.05026	2966.6	0.04904	2963.3	0.04787	2959.9

Table C.7 Specific volumes and specific enthalpies of steam at various temperatures and pressures—cont'd

T	v	h	v	h	v	h
°C	m³/kg	kJ/kg	m³/kg	kJ/kg	m³/kg	kJ/kg
320	0.05167	2995.4	0.05043	2992.4	0.04925	2989.3
330	0.05303	3023.2	0.05178	3020.4	0.05058	3017.5
340	0.05436	3050.1	0.05309	3047.5	0.05187	3044.9
350	0.05565	3076.5	0.05437	3074.1	0.05313	3071.6
360	0.05692	3102.7	0.05561	3100.5	0.05436	3098.2
370	0.05816	3128.1	0.05684	3126.0	0.05557	3123.9
380	0.05939	3153.2	0.05804	3151.2	0.05675	3149.2
390	0.06059	3177.9	0.05923	3176.0	0.05792	3174.1
400	0.06178	3202.2	0.06040	3200.5	0.05907	3198.7
410	0.06295	3226.4	0.06155	3224.7	0.06021	3223.0
420	0.06412	3250.3	0.06269	3248.7	0.06133	3247.0
430	0.06527	3274.0	0.06383	3272.4	0.06244	3270.9
440	0.06641	3297.6	0.06495	3296.1	0.06354	3294.6
450	0.06754	3321.0	0.06606	3319.6	0.06464	3318.2
460	0.06866	3344.4	0.06716	3343.1	0.06572	3341.7
470	0.06978	3367.7	0.06826	3366.4	0.06680	3365.1
480	0.07089	3390.9	0.06934	3389.7	0.06787	3388.4
490	0.07199	3414.1	0.07043	3412.9	0.06893	3411.7
500	0.07309	3437.2	0.07150	3436.1	0.06998	3434.9
510	0.07418	3460.3	0.07257	3459.2	0.07103	3458.1
520	0.07526	3483.4	0.07364	3482.3	0.07208	3481.2
530	0.07634	3506.4	0.07470	3505.4	0.07312	3504.4
540	0.07742	3529.5	0.07575	3528.5	0.07416	3527.5
550	0.07849	3552.5	0.07680	3551.5	0.07519	3550.6
560	0.07955	3575.6	0.07784	3574.6	0.07621	3573.7
570	0.08061	3598.6	0.07889	3597.7	0.07723	3596.8
580	0.08167	3621.7	0.07993	3620.8	0.07825	3620.0
590	0.08273	3644.7	0.08096	3643.9	0.07927	3643.1
600	0.08378	3667.8	0.08200	3667.0	0.08028	3666.2
610	0.08484	3690.8	0.08303	3690.0	0.08130	3689.2
620	0.08588	3713.9	0.08406	3713.1	0.08230	3712.3
630	0.08693	3736.9	0.08508	3736.2	0.08331	3735.4
640	0.08797	3760.0	0.08611	3759.3	0.08432	3758.6
650	0.08902	3783.2	0.08713	3782.5	0.08532	3781.8
	$p = 5.0$ MPa		$p = 5.2$ MPa		$p = 5.4$ MPa	
270	0.04053	2818.4	0.03847	2806.8	0.03654	2794.8
280	0.04224	2857.0	0.04018	2847.1	0.03825	2836.8
290	0.04383	2892.4	0.04175	2883.7	0.03981	2874.8

Continued

Table C.7 Specific volumes and specific enthalpies of steam at various temperatures and pressures—cont'd

T	υ	h	υ	h	υ	h
°C	m³/kg	kJ/kg	m³/kg	kJ/kg	m³/kg	kJ/kg
300	0.04532	2925.4	0.04322	2917.7	0.04127	2909.8
310	0.04675	2956.5	0.04462	2949.4	0.04265	2942.5
320	0.04811	2986.2	0.04596	2979.9	0.04397	2973.5
330	0.04942	3014.6	0.04725	3008.9	0.04523	3003.0
340	0.05070	3042.2	0.04849	3036.7	0.04645	3031.5
350	0.05194	3069.2	0.04971	3064.3	0.04763	3059.3
360	0.05316	3095.9	0.05089	3091.3	0.04878	3086.7
370	0.05435	3121.8	0.05204	3117.5	0.04991	3113.1
380	0.05551	3147.2	0.05318	3143.1	0.05102	3139.1
390	0.05666	3172.2	0.05430	3168.4	0.05210	3164.5
400	0.05780	3196.9	0.05541	3193.4	0.05317	3189.7
410	0.05891	3221.3	0.05648	3217.9	0.05422	3214.4
420	0.06002	3245.4	0.05755	3242.2	0.05526	3238.9
430	0.06111	3269.4	0.05861	3266.3	0.05629	3263.2
440	0.06220	3293.2	0.05966	3290.2	0.05731	3287.2
450	0.06327	3316.8	0.06070	3314.0	0.05831	3311.2
460	0.06434	3340.4	0.06173	3337.7	0.05931	3335.0
470	0.06539	3363.8	0.06275	3361.2	0.06030	3358.6
480	0.06644	3387.2	0.06377	3384.7	0.06129	3382.2
490	0.06749	3410.5	0.06478	3408.1	0.06226	3405.7
510	0.06853	3433.8	0.06578	3431.5	0.06323	3429.2
510	0.06956	3457.0	0.06677	3454.8	0.06420	3452.6
520	0.07058	3480.2	0.06777	3478.1	0.06516	3475.9
530	0.07161	3503.4	0.06875	3501.3	0.06611	3499.2
540	0.07262	3526.5	0.06974	3524.6	0.06706	3522.5
550	0.07363	3549.6	0.07071	3547.7	0.06800	3545.8
560	0.07464	3572.8	0.07168	3570.9	0.06894	3569.1
570	0.07564	3596.0	0.07265	3594.2	0.06988	3592.4
500	0.07665	3619.1	0.07362	3617.4	0.07081	3615.6
590	0.07764	3642.2	0.07458	3640.5	0.07174	3638.9
600	0.07864	3665.4	0.07554	3663.7	0.07267	3662.1
610	0.07963	3688.4	0.07650	3686.9	0.07359	3685.3
620	0.08062	3711.6	0.07745	3710.0	0.07452	3708.5
630	0.08161	3734.7	0.07840	3733.2	0.07543	3731.7
640	0.08259	3757.9	0.07935	3756.4	0.07635	3755.0
650	0.08358	3781.1	0.08030	3779.6	0.07727	3778.2
	$p = 5.6$ MPa		$p = 5.8$ MPa		$p = 6.0$ MPa	
280	0.03645	2826.3	0.03476	2815.3	0.03317	2804.0
290	0.03801	2865.6	0.03632	2856.2	0.03473	2846.5

Table C.7 Specific volumes and specific enthalpies of steam at various temperatures and pressures—cont'd

T	v	h	v	h	v	h
°C	m³/kg	kJ/kg	m³/kg	kJ/kg	m³/kg	kJ/kg
300	0.03946	2901.7	0.03776	2893.4	0.03616	2885.0
310	0.04082	2935.3	0.03910	2927.9	0.03750	2920.4
320	0.04211	2966.9	0.04038	2960.3	0.03876	2953.5
330	0.04335	2997.1	0.04160	2991.0	0.03996	2984.9
340	0.04454	3026.0	0.04277	3020.5	0.04111	3014.9
350	0.04570	3054.2	0.04391	3049.1	0.04223	3043.9
360	0.04683	3082.0	0.04501	3077.2	0.04331	3072.4
370	0.04793	3108.7	0.04608	3104.3	0.04436	3099.8
380	0.04901	3135.0	0.04713	3130.8	0.04538	3126.6
390	0.05006	3160.7	0.04816	3156.8	0.04639	3152.9
400	0.05110	3186.0	0.04918	3182.4	0.04738	3178.6
410	0.05213	3211.0	0.05018	3207.5	0.04835	3204.0
420	0.05314	3235.6	0.05116	3232.3	0.04931	3229.0
430	0.05414	3260.1	0.05213	3256.9	0.05026	3253.8
440	0.05512	3284.3	0.05309	3281.3	0.05119	3278.3
450	0.05610	3308.3	0.05404	3305.5	0.05212	3302.6
460	0.05707	3332.2	0.05498	3329.5	0.05303	3326.8
470	0.05803	3356.0	0.05591	3353.4	0.05394	3350.8
480	0.05898	3379.7	0.05684	3377.2	0.05484	3374.7
490	0.05993	3403.3	0.05776	3400.9	0.05573	3398.5
500	0.06087	3426.9	0.05897	3424.5	0.05662	3422.2
510	0.06180	3450.3	0.05958	3448.1	0.05750	3445.9
520	0.06273	3473.8	0.06048	3471.6	0.05837	3469.5
550	0.06549	3543.8	0.06315	3541.9	0.06096	3540.0
560	0.06640	3567.2	0.06403	3565.4	0.06182	3563.5
570	0.06730	3590.6	0.06491	3588.8	0.06267	3587.0
560	0.06821	3613.9	0.06578	3612.1	0.06352	3610.4
590	0.06110	3637.2	0.06665	3635.5	0.06436	3633.8
600	0.07000	3660.4	0.06752	3658.8	0.06521	3657.2
610	0.07090	3683.7	0.06839	3682.1	0.06604	3680.6
620	0.07175	3707.0	0.06925	3705.4	0.06688	3703.9
630	0.07268	3730.2	0.07011	3728.7	0.06772	3727.2
640	0.07356	3753.5	0.07097	3752.1	0.06855	3750.6
650	0.07441	3776.8	0.07182	3775.4	0.06938	3774.0
	$p = 6.2$ MPa		$p = 6.4$ MPa		$p = 6.6$ MPa	
280	0.03166	2792.1	0.03024	2779.8	−	−
290	0.03324	2836.5	0.03182	2826.2	0.03049	2815.5
300	0.03467	2876.3	0.03325	2867.4	0.03192	2858.2

Continued

Table C.7 Specific volumes and specific enthalpies of steam at various temperatures and pressures—cont'd

T	v	h	v	h	v	h
°C	m³/kg	kJ/kg	m³/kg	kJ/kg	m³/kg	kJ/kg
310	0.03599	2912.7	0.03457	2904.9	0.03324	2896.8
320	0.03724	2946.0	0.03581	2939.6	0.03446	2932.5
330	0.03842	2978.6	0.03698	2972.3	0.03562	2965.9
340	0.03956	3009.2	0.03810	3003.4	0.03673	2997.5
350	0.04065	3038.6	0.03917	3033.3	0.03779	3027.9
360	0.04171	3067.7	0.04022	3062.7	0.03881	3057.8
370	0.04274	3095.3	0.04123	3090.8	0.03980	3086.2
380	0.04375	3122.4	0.04221	3118.1	0.04077	3113.9
390	0.04473	3148.9	0.04318	3144.9	0.04171	3140.9
400	0.04570	3174.9	0.04412	3171.2	0.04263	3167.4
410	0.04665	3200.5	0.04505	3196.9	0.04354	3193.3
420	0.04758	3225.7	0.04596	3222.2	0.04444	3218.9
430	0.04850	3250.6	0.04686	3247.4	0.04532	3244.2
440	0.04941	3275.2	0.04775	3272.2	0.04618	3269.2
450	0.05032	3299.7	0.04863	3296.8	0.04704	3293.9
460	0.05121	3324.0	0.04950	3321.2	0.04789	3318.4
470	0.05209	3348.1	0.05036	3345.5	0.04873	3342.8
480	0.05296	3372.2	0.05121	3369.6	0.04956	3367.1
490	0.05383	3396.1	0.05205	3393.6	0.05038	3391.2
500	0.05469	3419.9	0.05289	3417.5	0.05120	3415.2
510	0.05555	3443.6	0.05372	3441.3	0.05201	3439.1
520	0.05640	3467.3	0.05455	3465.1	0.05282	3463.0
530	0.05725	3490.9	0.05537	3488.8	0.05362	3486.7
540	0.05809	3514.5	0.05619	3512.5	0.05441	3510.5
550	0.05892	3538.0	0.05700	3536.1	0.05520	3534.2
560	0.05975	3561.6	0.05781	3559.7	0.05599	3557.9
570	0.06058	3585.2	0.05862	3583.3	0.05677	3581.5
580	0.06140	3608.6	0.05942	3606.9	0.05755	3605.1
590	0.06222	3632.1	0.06021	3630.4	0.05833	3628.7
600	0.06304	3655.5	0.06101	3653.9	0.05910	3652.2
610	0.06385	3678.9	0.06180	3677.4	0.05987	3675.8
620	0.06466	3702.3	0.06259	3700.8	0.06064	3699.2
630	0.06547	3725.7	0.06337	3724.2	0.06140	3722.7
640	0.06628	3749.2	0.06416	3747.7	0.06216	3746.2
650	0.06708	3772.6	0.06494	3771.2	0.06292	3769.7
	p = 6.8 MPa		p = 7.0 MPa		p = 7.2 MPa	
290	0.02922	2804.4	0.02801	2792.9	0.02686	2780.9
300	0.03066	2848.8	0.02946	2839.2	0.02832	2829.0
310	0.03197	2888.6	0.03078	2880.2	0.02964	2871.6

Table C.7 Specific volumes and specific enthalpies of steam at various temperatures and pressures—cont'd

T	v	h	v	h	v	h
°C	m³/kg	kJ/kg	m³/kg	kJ/kg	m³/kg	kJ/kg
320	0.03319	2925.2	0.03199	2917.8	0.03085	2910.2
330	0.03434	2959.3	0.03313	2952.6	0.03198	2945.9
340	0.03543	2991.6	0.03421	2985.5	0.03305	2979.4
350	0.03647	3022.5	0.03524	3017.0	0.03407	3011.4
360	0.03748	3052.7	0.03623	3047.6	0.03505	3042.5
370	0.03846	3081.5	0.03719	3076.8	0.02599	3072.1
380	0.03941	3109.5	0.03812	3105.2	0.03691	3100.7
390	0.04033	3136.9	0.03903	3132.8	0.03780	3128.6
400	0.04124	3163.6	0.03992	3159.7	0.03867	3155.8
410	0.04213	3189.8	0.04079	3186.1	0.03953	3182.5
420	0.04300	3215.5	0.04165	3212.1	0.04036	3208.6
430	0.04386	3240.9	0.04249	3237.7	0.04119	3234.4
440	0.04471	3266.1	0.04332	3263.0	0.04201	3259.9
450	0.04555	3291.0	0.04414	3288.0	0.04281	3285.1
460	0.04637	3315.7	0.04495	3312.8	0.04360	3310.0
470	0.04719	3340.2	0.04575	3337.5	0.04438	3334.8
480	0.04800	3364.5	0.04654	3361.9	0.04516	3359.4
490	0.04881	3388.7	0.04732	3386.3	0.04592	3383.8
500	0.04961	3412.8	0.04810	3410.5	0.04668	3408.1
510	0.05040	3436.8	0.04888	3434.5	0.04744	3432.3
520	0.05118	3460.8	0.04964	3458.6	0.04819	3456.4
530	0.05196	3484.6	0.05040	3482.5	0.04893	3480.4
540	0.05274	3508.4	0.05116	3506.4	0.04967	3504.4
550	0.05351	3532.2	0.0519	3530.2	0.05040	3528.3
560	0.05427	3556.0	0.05266	3554.1	0.05113	3552.2
570	0.05504	3579.7	0.05340	3577.9	0.05186	3576.1
580	0.05580	3603.4	0.05414	3601.6	0.05258	3599.9
590	0.05655	3627.0	0.05488	3625.3	0.05330	3623.6
600	0.05730	3650.6	0.05561	3649.0	0.05401	3647.3
610	0.05805	3674.2	0.05634	3672.6	0.05472	3671.0
620	0.05880	3697.7	0.05707	3696.2	0.05543	3694.6
630	0.05954	3721.2	0.05779	3719.7	0.05614	3718.2
640	0.06028	3744.8	0.05851	3743.3	0.05684	3741.8
650	0.06102	3768.3	0.05923	3766.9	0.05754	3765.5
	$p = 7.4$ MPa		$p = 7.6$ MPa		$p = 7.8$ MPa	
290	0.02575	2768.4	—	—	—	—
300	0.02724	2818.7	0.02620	2808.0	0.02520	2796.9
310	0.02856	2862.7	0.02753	2853.6	0.02654	2844.3

Continued

Table C.7 Specific volumes and specific enthalpies of steam at various temperatures and pressures—cont'd

T	v	h	v	h	v	h
°C	m³/kg	kJ/kg	m³/kg	kJ/kg	m³/kg	kJ/kg
320	0.02977	2902.4	0.02874	2894.5	0.02775	2886.4
330	0.03089	2939.0	0.02985	2931.9	0.02887	2924.8
340	0.03195	2973.2	0.03091	2966.9	0.02991	2960.4
350	0.03296	3005.8	0.03190	3000.0	0.03090	2994.2
360	0.03393	3037.3	0.03286	3032.0	0.03185	3026.7
370	0.03486	3067.3	0.03378	3062.4	0.03276	3057.5
380	0.03576	3096.3	0.03467	3091.8	0.03364	3087.3
390	0.03664	3124.5	0.03554	3120.3	0.03449	3116.1
400	0.03750	3152.0	0.03638	3148.0	0.03532	3144.1
410	0.03834	3178.8	0.03720	3175.1	0.03613	3171.4
420	0.03916	3205.2	0.03801	3201.7	0.03692	3198.2
430	0.03997	3231.2	0.03880	3227.8	0.03770	3224.5
440	0.04076	3256.8	0.03959	3253.6	0.03847	3250.5
450	0.04155	3282.1	0.04036	3279.1	0.03922	3276.1
460	0.04232	3307.2	0.04112	3304.4	0.03997	3301.5
470	0.04309	3332.1	0.04187	3329.4	0.04070	3326.6
480	0.04385	3356.8	0.0426!	3354.2	0.04143	3351.6
490	0.04460	3381.3	0.04334	3378.8	0.04215	3376.3
500	0.04534	3405.7	0.04407	3403.3	0.04286	3400.9
510	0.04608	3430.0	0.04479	3427.7	0.04357	3425.4
520	0.04681	3454.2	0.04551	3452.0	0.04427	3449.8
530	0.04754	3478.3	0.04622	3476.2	0.04496	3474.1
540	0.04826	3502.3	0.04692	3500.3	0.04565	3498.3
550	0.04898	3526.3	0.04762	3524.4	0.04634	3522.4
560	0.04969	3550.3	0.04832	3548.4	0.04702	3546.5
570	0.05039	3574.2	0.04901	3572.4	0.04770	3570.6
580	0.05110	3598.1	0.04970	3596.3	0.04837	3594.6
590	0.05180	3621.9	0.05038	3620.2	0.04904	3618.5
600	0.05250	3645.6	0.05106	3644.0	0.04970	3642.3
610	0.05319	3669.4	0.05174	3667.8	0.05037	3666.2
620	0.05388	3693.0	0.05242	3691.5	0.05103	3689.9
630	0.05457	3716.7	0.05309	3715.2	0.05168	3713.7
640	0.05526	3740.3	0.05376	3738.9	0.05234	3737.4
650	0.05594	3764.0	0.05443	3762.6	0.05299	3761.2
	$p = 8.0$ MPa		$p = 8.2$ MPa		$p = 8.4$ MPa	
300	0.02425	2785.4	0.02333	2773.5	0.02244	2760.9
310	0.02560	2834.7	0.02470	2824.8	0.02383	2814.5
320	0.02682	2878.1	0.02592	2869.6	0.02506	2860.9
330	0.02793	2917.5	0.02703	2910.0	0.02617	2902.5

Table C.7 Specific volumes and specific enthalpies of steam at various temperatures and pressures—cont'd

T	υ	h	υ	h	υ	h
°C	m³/kg	kJ/kg	m³/kg	kJ/kg	m³/kg	kJ/kg
340	0.02897	2953.9	0.02807	2947.3	0.02721	2940.6
350	0.02995	2988.3	0.02904	2982.3	0.02818	2976.2
360	0.03089	3021.3	0.02997	3015.8	0.02910	3010.3
370	0.03179	3052.6	0.03086	3047.6	0.02998	3042.5
380	0.03265	3082.7	0.03171	3078.1	0.03082	3073.4
390	0.03349	3111.8	0.03254	3107.5	0.03164	3103.2
400	0.03431	3140.1	0.03335	3136.1	0.03243	3132.0
410	0.03511	3167.7	0.03413	3163.9	0.03321	3160.1
420	0.03589	3194.7	0.03490	3191.1	0.03396	3187.5
430	0.03665	3221.2	0.03566	3217.8	0.03470	3214.4
440	0.03741	3247.3	0.03640	3244.1	0.03543	3240.9
450	0.03815	3273.1	0.03712	3270.1	0.03615	3267.1
460	0.03888	3298.6	0.03784	3295.8	0.03685	3292.9
470	0.03960	3323.9	0.03855	3321.2	0.03755	3318.4
480	0.04031	3349.0	0.03925	3346.3	0.03824	3343.7
490	0.04102	3373.8	0.03994	3371.3	0.03892	3368.8
500	0.04172	3398.5	0.04062	3396.1	0.03959	3393.7
510	0.04241	3423.1	0.04130	3420.8	0.04025	3418.5
520	0.04309	3447.6	0.04197	3445.4	0.04091	3443.1
530	0.04377	3471.9	0.04264	3469.8	0.04156	3467.7
540	0.04445	3496.2	0.04330	3494.2	0.04221	3492.1
550	0.04512	3520.4	0.04396	3518.4	0.04286	3516.4
560	0.04578	3544.6	0.04461	3542.7	0.04350	3540.8
570	0.04645	3568.7	0.04526	3566.9	0.04413	3565.0
580	0.04710	3592.8	0.04590	3591.0	0.04476	3589.2
590	0.04776	3616.8	0.04654	3615.0	0.04539	3613.3
600	0.04841	3640.7	0.04718	3639.0	0.04601	3637.4
610	0.04906	3664.6	0.04782	3662.9	0.04663	3661.3
620	0.04970	3688.4	0.04845	3686.8	0.04725	3685.3
630	0.05035	3712.2	0.04906	3710.7	0.04786	3709.2
640	0.05099	3736.0	0.04970	3734.5	0.04848	3733.0
650	0.05162	3759.8	0.05032	3758.4	0.04909	3756.9
	$p = 8.6$ MPa		$p = 8.8$ MPa		$p = 9.0$ MPa	
310	0.02300	2804.0	0.02219	2793.1	0.02142	2781.8
320	0.02424	2852.0	0.02344	2842.9	0.02268	2833.5
330	0.02535	2894.7	0.02456	2886.8	0.02381	2878.7
340	0.02638	2933.7	0.02559	2926.7	0.02484	2919.6
350	0.02735	2970.1	0.02655	2963.8	0.02579	2957.5

Continued

Table C.7 Specific volumes and specific enthalpies of steam at various temperatures and pressures—cont'd

T	υ	h	υ	h	υ	h
°C	m³/kg	kJ/kg	m³/kg	kJ/kg	m³/kg	kJ/kg
360	0.02826	3004.7	0.02746	2999.0	0.02669	2993.2
370	0.02913	3037.4	0.02832	3032.2	0.02755	3027.0
380	0.02998	3068.7	0.02915	3063.9	0.02837	3059.1
390	0.03078	3098.8	0.02995	3094.4	0.02916	3090.0
400	0.03156	3127.9	0.03072	3123.8	0.02993	3119.7
410	0.03232	3156.3	0.03148	3152.4	0.03067	3148.5
420	0.03307	3183.9	0.03221	3180.3	0.03139	3176.7
430	0.03380	3211.1	0.03293	3207.6	0.03210	3204.2
440	0.03451	3237.7	0.03364	3234.5	0.03280	3231.2
450	0.03522	3264.0	0.03433	3261.0	0.03348	3257.9
460	0.03591	3290.0	0.03501	3287.1	0.03415	3284.1
470	0.03659	3315.6	0.03568	3312.9	0.03481	3310.1
480	0.03727	3341.1	0.03635	3338.4	0.03546	3335.7
490	0.03794	3366.3	0.03700	3363.7	0.03611	3361.2
500	0.03860	3391.3	0.03765	3388.9	0.03675	3386.4
510	0.03925	3416.2	0.03829	3413.8	0.03738	3411.5
520	0.03990	3440.9	0.03893	3438.7	0.03800	3436.4
530	0.04054	3465.5	0.03955	3463.4	0.03862	3461.2
540	0.04117	3490.0	0.04018	3488.0	0.03923	3485.9
550	0.04180	3514.5	0.04080	3512.5	0.03984	3510.5
560	0.04243	3538.9	0.04141	3537.0	0.04044	3535.0
570	0.04305	3563.2	0.04202	3561.4	0.04104	3559.5
580	0.04367	3587.4	0.04263	3585.7	0.04163	3583.9
590	0.04428	3611.6	0.04323	3609.9	0.04222	3608.2
600	0.04490	3635.7	0.04383	3634.0	0.04281	3632.4
610	0.04550	3659.7	0.04443	3658.1	0.04340	3656.5
620	0.04611	3683.7	0.04502	3682.1	0.04398	3680.6
630	0.04671	3707.6	0.04561	3706.1	0.04456	3704.6
640	0.04731	3731.6	0.04620	3730.1	0.04513	3728.6
650	0.04791	3755.5	0.04678	3754.1	0.04571	3752.6
	$p = 9.2$ MPa		$p = 9.4$ MPa		$p = 9.6$ MPa	
310	0.02067	2770.1	0.01994	2757.7	0.01923	2745.0
320	0.02195	2823.8	0.02124	2813.8	0.02055	2803.6
330	0.02308	2870.5	0.02238	2862.0	0.02170	2853.4
340	0.02411	2912.4	0.02341	2905.0	0.02274	2897.5
350	0.02506	2951.0	0.02436	2944.5	0.02369	2937.9
360	0.02596	2987.4	0.02526	2981.5	0.02458	2975.5
370	0.02681	3021.7	0.02610	3016.3	0.02542	3010.9
380	0.02762	3054.3	0.02691	3049.4	0.02622	3044.4

Table C.7 Specific volumes and specific enthalpies of steam at various temperatures and pressures—cont'd

T	υ	h	υ	h	υ	h
°C	m³/kg	kJ/kg	m³/kg	kJ/kg	m³/kg	kJ/kg
390	0.02841	3085.5	0.02768	3080.9	0.02699	3076.4
400	0.02916	3115.5	0.02843	3111.3	0.02773	3107.1
410	0.02990	3144.6	0.02915	3140.7	0.02845	3136.8
420	0.03061	3173.0	0.02986	3169.3	0.02914	3165.6
430	0.03131	3200.8	0.03055	3197.3	0.02982	3193.8
440	0.03200	3228.0	0.03123	3224.7	0.03049	3221.4
450	0.03267	3254.8	0.03189	3251.7	0.03114	3248.5
460	0.03333	3281.2	0.03254	3278.2	0.03178	3275.3
470	0.03398	3307.3	0.03318	3304.5	0.03242	3301.6
480	0.03462	3333.1	0.03381	3330.4	0.03304	3327.7
490	0.03525	3358.6	0.03444	3356.1	0.03365	3353.5
500	0.03588	3384.0	0.03505	3381.5	0.03426	3379.1
510	0.03650	3409.2	0.03566	3406.8	0.03486	3404.5
520	0.03711	3434.2	0.03626	3431.9	0.03545	3429.7
530	0.03772	3459.0	0.03686	3456.9	0.03604	3454.7
540	0.03832	3483.8	0.03745	3481.7	0.03662	3479.6
550	0.03892	3508.5	0.03804	3506.5	0.03720	3504.4
560	0.03951	3533.1	0.03862	3531.2	0.03777	3529.2
570	0.04010	3557.6	0.03920	3555.8	0.03833	3553.9
580	0.04068	3582.1	0.03977	3580.3	0.03890	3578.5
590	0.04126	3606.4	0.04034	3604.7	0.03946	3603.0
600	0.04184	3630.7	0.04091	3629.0	0.04001	3627.3
610	0.04241	3654.9	0.04147	3653.2	0.04057	3651.6
620	0.04298	3679.0	0.04203	3677.4	0.04112	3675.9
630	0.04355	3703.1	0.04259	3701.6	0.04166	3700.0
640	0.04412	3727.1	0.04314	3725.7	0.04221	3724.2
650	0.04468	3751.2	0.04369	3749.8	0.04275	3748.3
	$p = 9.8$ MPa		$p = 10.0$ MPa		$p = 10.5$ MPa	
310	0.01853	2731.7	—	—	—	—
320	0.01989	2793.0	0.01924	2782.0	0.01771	2752.8
330	0.02105	2844.5	0.02042	2835.4	0.01893	2811.6
340	0.02209	2889.9	0.02147	2882.1	0.02000	2861.8
350	0.02304	2931.1	0.02242	2924.2	0.02095	2906.5
360	0.02393	2969.5	0.02330	2963.3	0.02184	2947.6
370	0.02477	3005.4	0.02414	2999.9	0.02266	2985.7
380	0.02556	3039.4	0.02492	3034.4	0.02344	3021.5
390	0.02632	3071.8	0.02568	3067.1	0.02418	3055.3
400	0.02705	3102.8	0.02641	3098.5	0.02489	3087.6

Continued

Table C.7 Specific volumes and specific enthalpies of steam at various temperatures and pressures—cont'd

T	v	h	v	h	v	h
°C	m³/kg	kJ/kg	m³/kg	kJ/kg	m³/kg	kJ/kg
410	0.02776	3132.8	0.02711	3128.7	0.02558	3118.6
420	0.02845	3161.9	0.02779	3158.1	0.02624	3148.6
430	0.02912	3190.3	0.02845	3186.7	0.02689	3177.8
440	0.02978	3218.0	0.02910	3214.8	0.02752	3206.4
450	0.03043	3245.4	0.02974	3242.2	0.02814	3234.3
460	0.03106	3272.3	0.03036	3269.3	0.02874	3261.8
470	0.03168	3298.8	0.03098	3296.0	0.02933	3288.8
480	0.03230	3325.0	0.03158	3322.3	0.02992	3315.5
490	0.03290	3350.9	0.03218	3348.3	0.03049	3341.8
500	0.03350	3376.6	0.03277	3374.1	0.03106	3367.9
510	0.03409	3402.1	0.03335	3399.7	0.03162	3393.8
520	0.03467	3427.4	0.03392	3425.1	0.03217	3419.4
530	0.03525	3452.5	0.03449	3450.3	0.03272	3444.9
540	0.03582	3477.5	0.03505	3475.4	0.03326	3470.2
550	0.03639	3502.4	0.03561	3500.4	0.03380	3495.4
560	0.03695	3527.3	0.03616	3525.4	0.03433	3520.5
570	0.03751	3552.0	0.03671	3550.2	0.03486	3545.5
580	0.03806	3576.7	0.03726	3574.9	0.03538	3570.4
590	0.03861	3601.2	0.03780	3599.5	0.03590	3595.1
600	0.03916	3625.6	0.03833	3624.0	0.03641	3619.8
610	0.03397	3650.0	0.03887	3648.4	0.03692	3644.3
620	0.04024	3674.3	0.03940	3672.7	0.03743	3668.8
630	0.04078	3698.5	0.03992	3697.0	0.03794	3693.2
640	0.04131	3722.7	0.04045	3721.2	0.03844	3717.5
650	0.04184	3746.9	0.04097	3745.4	0.03894	3741.8
	$p = 11.0$ MPa		$p = 11.5$ MPa		$p = 12.0$ MPa	
320	0.01625	2720.3	—	—	—	—
330	0.01754	2785.9	0.01624	2758.1	0.01501	2727.6
340	0.01864	2840.4	0.01738	2817.7	0.01620	2793.4
350	0.01961	2888.1	0.01836	2868.7	0.01721	2848.4
360	0.02049	2931.2	0.01925	2914.3	0.01810	2896.6
370	0.02131	2971.1	0.02007	2956.0	0.01893	2940.4
380	0.02208	3008.3	0.02084	2994.7	0.01969	2980.7
390	0.02281	3043.2	0.02156	3030.8	0.02040	3018.1
400	0.02351	3076.4	0.02224	3065.0	0.02108	3053.3
410	0.02418	3108.2	0.02290	3097.7	0.02173	3086.9
420	0.02483	3138.9	0.02354	3129.1	0.02235	3119.1
430	0.02546	3168.8	0.02416	3159.6	0.02296	3150.2
440	0.02607	3197.8	0.02475	3189.2	0.02354	3180.4

Table C.7 Specific volumes and specific enthalpies of steam at various temperatures and pressures—cont'd

T	υ	h	υ	h	υ	h
°C	m³/kg	kJ/kg	m³/kg	kJ/kg	m³/kg	kJ/kg
450	0.02667	3226.2	0.02534	3218.1	0.02411	3209.9
460	0.02726	3254.1	0.02591	3246.4	0.02467	3238.6
470	0.02784	3281.6	0.02647	3274.2	0.02521	3266.9
480	0.02840	3308.6	0.02702	3301.6	0.02575	3294.6
490	0.02896	3335.3	0.02756	3328.6	0.02627	3322.0
500	0.02951	3361.6	0.02809	3355.3	0.02679	3349.0
510	0.03005	3387.8	0.02861	3381.7	0.02730	3375.6
520	0.03058	3413.7	0.02913	3407.9	0.02780	3402.1
530	0.03111	3439.4	0.02964	3433.8	0.02829	3428.2
540	0.03163	3464.9	0.03014	3459.6	0.02878	3454.2
550	0.03215	3490.3	0.03064	3485.2	0.02926	3480.0
560	0.03266	3515.6	0.03114	3510.7	0.02974	3505.7
570	0.03317	3540.8	0.03163	3536.0	0.03022	3531.3
580	0.03367	3565.8	0.03211	3561.3	0.03068	3556.7
590	0.03417	3590.7	0.03260	3586.3	0.03115	3581.9
600	0.03467	3615.5	0.03307	3611.3	0.03161	3607.0
610	0.03516	3640.2	0.03355	3636.1	0.03207	3632.0
620	0.03565	3664.8	0.03402	3660.8	0.03252	3656.9
630	0.03614	3689.3	0.03449	3685.5	0.03298	3681.6
640	0.03662	3713.8	0.03495	3710.1	0.03343	3706.4
650	0.03710	3738.2	0.03542	3734.6	0.03387	3731.0
	$p = 12.5$ MPa		$p = 13.0$ MPa		$p = 13.5$ MPa	
330	0.01382	2693.4	—	—	—	—
340	0.01508	2767.2	0.01402	2738.8	0.01300	2707.6
350	0.01613	2826.8	0.01511	2804.0	0.01415	2779.7
360	0.01704	2878.2	0.01604	2858.9	0.01511	2838.7
370	0.01786	2924.3	0.01688	2907.5	0.01595	2890.1
380	0.01862	2966.2	0.01764	2951.4	0.01672	2936.0
390	0.01933	3005.0	0.01834	2991.7	0.01742	2977.9
400	0.02001	3041.4	0.01901	3029.3	0.01808	3016.8
410	0.02065	3075.9	0.01964	3064.8	0.01871	3053.4
420	0.02126	3109.0	0.02025	3098.6	0.01931	3088.1
430	0.02185	3140.8	0.02083	3131.1	0.01988	3121.4
440	0.02243	3171.5	0.02139	3162.6	0.02044	3153.4
450	0.02298	3201.5	0.02194	3193.1	0.02097	3184.5
460	0.02353	3230.8	0.02247	3222.8	0.02149	3214.7
470	0.02406	3259.4	0.02299	3251.9	0.02200	3244.3
480	0.02458	3287.5	0.02350	3280.4	0.02250	3273.2

Continued

Table C.7 Specific volumes and specific enthalpies of steam at various temperatures and pressures—cont'd

T	v	h	v	h	v	h
°C	m³/kg	kJ/kg	m³/kg	kJ/kg	m³/kg	kJ/kg
490	0.02509	3315.2	0.02400	3308.5	0.02298	3301.6
500	0.02559	3342.6	0.02448	3336.1	0.02346	3329.6
510	0.02608	3369.5	0.02496	3363.4	0.02393	3357.2
520	0.02657	3396.2	0.02544	3390.3	0.02439	3384.4
530	0.02705	3422.6	0.02590	3417.0	0.02484	3411.3
540	0.02752	3448.8	0.02637	3443.4	0.02529	3438.0
550	0.02799	3474.9	0.02682	3469.7	0.02574	3464.5
560	0.02846	3500.8	0.02727	3495.8	0.02617	3490.8
570	0.02892	3526.5	0.02772	3521.7	0.02661	3516.9
580	0.02937	3552.1	0.02816	3547.5	0.02703	3542.8
590	0.02982	3577.5	0.02859	3573.0	0.02746	3568.6
600	0.03027	3602.7	0.02903	3598.4	0.02788	3594.2
610	0.03071	3627.9	0.02946	3623.7	0.02829	3619.6
620	0.03115	3652.9	0.02988	3648.9	0.02871	3644.9
630	0.03159	3677.8	0.03030	3673.9	0.02912	3670.0
640	0.03202	3702.6	0.03072	3698.9	0.02952	3695.1
650	0.03245	3727.4	0.03114	3723.8	0.02993	3720.1
	$p = 14.0$ MPa		$p = 14.5$ MPa		$p = 15.0$ MPa	
340	0.01201	2672.6	0.01101	2632.3	—	—
350	0.01323	2753.5	0.01235	2725.1	0.01148	2693.8
360	0.01422	2817.4	0.01338	2794.9	0.01258	2771.3
370	0.01508	2871.9	0.01426	2853.0	0.01349	2833.3
380	0.01585	2920.2	0.01504	2903.7	0.01428	2886.7
390	0.01656	2963.8	0.01575	2949.3	0.01500	2934.3
400	0.01722	3004.0	0.01641	2991.0	0.01566	2977.6
410	0.01784	3041.7	0.01703	3029.8	0.01627	3017.7
420	0.01844	3077.4	0.01762	3066.4	0.01685	3055.3
430	0.01900	3111.4	0.01818	3101.3	0.01741	3091.1
440	0.01954	3144.2	0.01872	3134.8	0.01794	3125.2
450	0.02007	3175.8	0.01923	3167.0	0.01845	3158.2
460	0.02058	3206.6	0.01974	3198.3	0.01894	3190.0
470	0.02108	3236.6	0.02022	3228.8	0.01942	3221.0
480	0.02157	3265.9	0.02070	3258.6	0.01989	3251.2
490	0.02204	3294.7	0.02117	3287.8	0.02035	3280.7
500	0.02251	3323.0	0.02162	3316.4	0.02079	3309.7
510	0.02297	3350.9	0.02207	3344.6	0.02123	3338.3
520	0.02342	3378.4	0.02251	3372.5	0.02166	3366.4
530	0.02386	3405.6	0.02294	3399.9	0.02206	3394.1
540	0.02430	3432.5	0.02337	3427.0	0.02250	3421.5

Table C.7 Specific volumes and specific enthalpies of steam at various temperatures and pressures—cont'd

T	v	h	v	h	v	h
°C	m³/kg	kJ/kg	m³/kg	kJ/kg	m³/kg	kJ/kg
550	0.02473	3459.2	0.02379	3454.0	0.02291	3448.7
560	0.02515	3485.8	0.02420	3480.7	0.02332	3475.6
570	0.02557	3512.1	0.02461	3507.2	0.02372	3502.3
580	0.02599	3538.2	0.02502	3533.5	0.02411	3528.8
590	0.02640	3564.1	0.02542	3559.6	0.02450	3555.1
600	0.02681	3589.8	0.02582	3585.5	0.02489	3581.2
610	0.02722	3615.4	0.02621	3611.2	0.02527	3607.0
620	0.02762	3640.8	0.02660	3636.8	0.02565	3632.8
630	0.02801	3666.1	0.02699	3662.2	0.02603	3658.3
640	0.02841	3691.3	0.02737	3687.6	0.02640	3683.8
650	0.02880	3716.5	0.02775	3712.8	0.02677	3709.1
	$p = 15.5$ MPa		$p = 16.0$ MPa		$p = 16.5$ MPa	
350	0.01064	2658.8	0.00978	2618.5	0.00889	2570.1
360	0.01181	2745.6	0.01107	2717.8	0.01034	2687.4
370	0.01275	2812.4	0.01205	2790.5	0.01137	2767.2
380	0.01356	2869.0	0.01286	2850.6	0.01222	2831.3
390	0.01428	2918.8	0.01360	2902.9	0.01296	2886.4
400	0.01494	2963.8	0.01427	2949.7	0.01363	2935.2
410	0.01556	3005.2	0.01488	2992.5	0.01425	2979.6
420	0.01614	3044.0	0.01546	3032.5	0.01482	3020.7
430	0.01668	3080.6	0.01600	3070.0	0.01536	3059.2
440	0.01721	3115.6	0.01652	3105.8	0.01588	3095.8
450	0.01771	3149.2	0.01702	3140.0	0.01637	3130.8
460	0.01820	3181.6	0.01750	3173.0	0.01685	3164.4
470	0.01867	3213.0	0.01797	3205.0	0.01731	3196.9
480	0.01913	3243.7	0.01842	3236.2	0.01775	3228.5
490	0.01958	3273.7	0.01886	3266.5	0.01819	3259.3
500	0.02002	3303.0	0.01929	3296.3	0.01861	3289.5
510	0.02045	3331.9	0.01971	3325.4	0.01902	3319.0
520	0.02087	3360.3	0.02013	3354.2	0.01943	3348.0
530	0.02128	3388.3	0.02053	3382.4	0.01982	3376.6
540	0.02169	3416.0	0.02093	3410.4	0.02021	3404.8
550	0.02209	3443.4	0.02132	3438.0	0.02060	3432.6
560	0.02249	3470.5	0.02171	3465.4	0.02098	3460.2
570	0.02288	3497.4	0.02209	3492.5	0.02135	3487.6
580	0.02326	3524.1	0.02247	3519.4	0.02172	3514.6
590	0.02364	3550.6	0.02284	3546.0	0.02208	3541.5
600	0.02402	3576.8	0.02321	3572.4	0.02244	3568.0

Continued

Table C.7 Specific volumes and specific enthalpies of steam at various temperatures and pressures—cont'd

T	v	h	v	h	v	h
°C	m³/kg	kJ/kg	m³/kg	kJ/kg	m³/kg	kJ/kg
610	0.02440	3602.8	0.02357	3598.6	0.02280	3594.4
620	0.02477	3628.7	0.02393	3624.6	0.02315	3620.6
630	0.02513	3654.4	0.02429	3650.5	0.02350	3646.5
640	0.02550	3680.0	0.02465	3676.2	0.02385	3672.4
650	0.02586	3705.5	0.02500	3701.8	0.02419	3698.1
	$p = 17.0$ MPa		$p = 17.5$ MPa		$p = 18.0$ MPa	
360	0.00962	2653.6	0.00889	2615.2	0.00814	2569.8
370	0.01072	2742.3	0.01008	2715.7	0.00946	2686.8
380	0.01160	2811.2	0.01100	2790.1	0.01042	2767.8
390	0.01235	2869.3	0.01177	2851.5	0.01122	2833.1
400	0.01303	2920.2	0.01246	2904.9	0.01191	2889.0
410	0.01365	2966.3	0.01308	2952.6	0.01254	2938.7
420	0.01422	3008.6	0.01365	2996.4	0.01311	2983.9
430	0.01476	3048.3	0.01419	3037.1	0.01365	3025.7
440	0.01527	3085.7	0.01470	3075.4	0.01415	3065.0
450	0.01576	3121.4	0.01518	3111.9	0.01463	3102.3
460	0.01623	3155.7	0.01565	3146.8	0.01509	3137.9
470	0.01668	3188.8	0.01610	3180.5	0.01554	3172.1
480	0.01712	3220.8	0.01653	3213.1	0.01597	3205.2
490	0.01755	3252.1	0.01695	3244.7	0.01638	3237.3
500	0.01797	3282.6	0.01736	3275.7	0.01678	3268.7
510	0.01837	3312.4	0.01776	3305.9	0.01718	3299.3
520	0.01877	3341.8	0.01815	3335.5	0.01756	3329.3
530	0.01916	3370.6	0.01853	3364.7	0.01794	3358.7
540	0.01954	3399.1	0.01891	3393.4	0.01831	3387.7
550	0.01992	3427.2	0.01928	3421.8	0.01867	3416.4
560	0.02029	3455.1	0.01964	3449.8	0.01903	3444.7
570	0.02066	3482.6	0.02000	3477.6	0.01938	3472.6
580	0.02102	3509.9	0.02035	3505.1	0.01973	3500.3
590	0.02137	3536.9	0.02070	3532.3	0.02007	3527.7
600	0.02173	3563.6	0.02105	3559.2	0.02041	3554.8
610	0.02207	3590.2	0.02139	3585.9	0.02074	3581.6
620	0.02242	3616.5	0.02172	3612.4	0.02107	3608.2
630	0.02276	3642.6	0.02206	3638.6	0.02140	3634.9
640	0.02310	3668.6	0.02239	3664.8	0.02172	3660.9
650	0.02343	3694.4	0.02272	3690.7	0.02204	3687.0
	$p = 18.5$ MPa		$p = 19.0$ MPa		$p = 19.5$ MPa	
360	0.00730	2511.7	—	—	—	—
370	0.00885	2655.1	0.00824	2619.8	0.00761	2579.4

Table C.7 Specific volumes and specific enthalpies of steam at various temperatures and pressures—cont'd

T	υ	h	υ	h	υ	h
°C	m³/kg	kJ/kg	m³/kg	kJ/kg	m³/kg	kJ/kg
380	0.00987	2744.2	0.00933	2719.1	0.00880	2692.3
390	0.01068	2813.9	0.01017	2793.9	0.00968	2772.9
400	0.01139	2872.7	0.01089	2855.7	0.01041	2838.2
410	0.01202	2924.3	0.01153	2909.5	0.01106	2894.5
420	0.01260	2971.1	0.01211	2958.0	0.01164	2944.6
430	0.01313	3014.2	0.01264	3002.4	0.01218	2990.4
440	0.01364	3054.4	0.01314	3043.7	0.01268	3032.8
450	0.01411	3092.5	0.01362	3082.6	0.01315	3072.6
460	0.01457	3128.8	0.01407	3119.7	0.01360	3110.4
470	0.01501	3163.7	0.01451	3155.1	0.01403	3146.5
480	0.01543	3197.3	0.01493	3189.3	0.01445	3181.2
490	0.01584	3229.9	0.01533	3222.4	0.01485	3214.8
500	0.01624	3261.7	0.01573	3254.5	0.01524	3247.7
510	0.01663	3292.6	0.01611	3285.9	0.01561	3279.1
520	0.01701	3322.9	0.01648	3316.6	0.01598	3310.2
530	0.01738	3352.7	0.01685	3346.7	0.01634	3340.6
540	0.01774	3382.0	0.01720	3376.2	0.01669	3370.4
550	0.01811	3410.9	0.01756	3405.4	0.01704	3399.9
560	0.01845	3439.4	0.01790	3434.2	0.01738	3428.9
570	0.01880	3467.6	0.01824	3462.6	0.01771	3457.5
580	0.01914	3495.5	0.01857	3490.6	0.01804	3485.8
590	0.01947	3523.0	0.01890	3518.4	0.01836	3513.8
600	0.01980	3550.3	0.01923	3545.9	0.01868	3541.4
610	0.02013	3577.4	0.01955	3573.1	0.01900	3568.8
620	0.02045	3604.1	0.01986	3600.0	0.01931	3595.8
630	0.02077	3630.7	0.02018	3626.7	0.01962	3622.7
640	0.02109	3657.1	0.02049	3653.2	0.01992	3649.4
650	0.02140	3683.3	0.02080	3679.6	0.02022	3675.9

REFERENCES

[1] Pis'mennyi EN. An asymptotic approach to generalizing the experimental data on convective heat transfer of tube bundles in crossflow. Int J Heat Mass Transfer 2011;54(19−20):4235−46.

[2] Pis'mennyi EN. The effectiveness of disk finning in tubular convective heat exchangers under cross-flow conditions. Therm Eng 1996;43(9):749−53.

[3] Pis'mennyi EN, Terekh AM, Polupan GP, Carvajal Mariscal I, Sánchez Silva F. Universal relations for calculation of the drag of transversely finned tube bundles. Int J Heat Mass Transfer 2014;73:293−302.

[4] Portyanko AA, Lokshin VA, Fomina VA, et al. Experimental investigations of tube bundles exposed to a transverse flow of gas with solid particles. Therm Eng (Теплоэнергетика) 1980;(6):14−8 [in Russian].

[5] Fomina VN, Titova EY, Migay VK, Bystrov PG, Pis'mennyi EN. Generalization of experimental data and development of recommendations for calculating heat transfer in staggered tube bundles with helical and circular fins exposed to a transverse flow of gas. Electric Stations (электрические станции) 1991;(6):48−56 [in Russian].

[6] Migay VK, Bystrov PG, Pis'mennyi EN, Zoz VN. Generalization of experimental data on convective heat transfer and aerodynamic resistance in tube bundles with circular fins. Transactions of TsKTI (Труды ЦКТИ), Leningrad, Russia 1987;(236):34−43 [in Russian].

[7] Pis'mennyi EN. Calculation of heat transfer and aerodynamics in tube bundles with circular fins. Kiev, Ukraine: Alterpres Publishing House; 2003. 184 pages [in Russian].

[8] Lokshin VA, Peterson DF, Schwarz AL, editors. Standard methods of hydraulic design for power boilers. Washington, DC, New York, NY, USA: Hemisphere Publishing Corporation; 1988. 345 pages.

[9] Lokshin VA, Chakrygin MP, Chebulaev MM, et al. On calculation of hydraulic schemes of steam overheaters. Therm Eng (Теплоэнергетика) 1978;(1):81−3 [in Russian].

[10] Kirilov PL, Yuriev YS, Bobkov VP. Augmented and Revised. In: Kirilov PL, Yuriev YS, Bobkov VP, editors. Thermohydraulic calculations handbook (nuclear reactors). 2nd ed. Moscow, Russia: Energoatomizdat Publishing House; 1990. 360 pages (in Russian).

[11] Vasserman AA, Kazavchinskiy YZ, Rabinovich VA. Thermophysical properties of air and its components. Moscow, Russia: Nauka Publishing House; 1966. 376 pages [in Russian].

[12] Standard mechanical calculation of power steam generators, steam and hot water piping. 1999. RD 10-249-98, Moscow, Russia, 228 pages [in Russian].

[13] Pysmennyy Ye, Polupan G, Carvajal Mariscal I, Sánchez Silva F. Handbook for heat exchangers and finned tube bundles calculation. Mexico City, Mexico: Reverté Publishing House; 2007. 197 pages [in Spanish].

INDEX

'*Note*: Page numbers followed by "f" indicate figures, "t" indicate tables.'